城市更新系列丛书

Construction Guidance for Renovation of Old Communities in Jiangsu

江苏老旧小区改造
建设导引

梅耀林　王承华　李琳琳　汤　蕾　樊思嘉　等 编著

中国建筑工业出版社

图书在版编目（CIP）数据

江苏老旧小区改造建设导引 = Construction Guidance for Renovation of Old Communities in Jiangsu / 梅耀林等编著. —北京：中国建筑工业出版社，2021.6
（城市更新系列丛书）
ISBN 978-7-112-26292-2

Ⅰ.①江… Ⅱ.①梅… Ⅲ.①居住区—旧房改造—研究—江苏 Ⅳ.①TU984.12

中国版本图书馆CIP数据核字（2021）第129703号

责任编辑：宋 凯 朱晓瑜
责任校对：芦欣甜

城市更新系列丛书
江苏老旧小区改造建设导引
Construction Guidance for Renovation of Old Communities in Jiangsu
梅耀林 王承华 李琳琳 汤 蕾 樊思嘉 等 编著

*

中国建筑工业出版社出版、发行（北京海淀三里河路9号）
各地新华书店、建筑书店经销
逸品书装设计制版
北京富诚彩色印刷有限公司印刷

*

开本：880毫米×1230毫米 1/16 印张：13¼ 字数：391千字
2021年7月第一版 2021年7月第一次印刷
定价：**98.00**元
ISBN 978-7-112-26292-2
（37856）

主编寄语

中国城镇化率已经越过60%的门槛，步入城镇化发展的中后期。中共中央“十四五”规划纲要提出实施城市更新行动，标志着我国城市建设将由大规模增量开发转为存量提质改造和增量结构调整并重的新阶段，提升城市宜居水平、实现从“住有所居”到“住有宜居”成为城市发展的重要议题，全面推进城镇老旧小区改造成为各地重要的民生工程和发展工程。为更好地指导老旧小区改造工作高质量开展，江苏省住房和城乡建设厅委托江苏省规划设计集团开展江苏老旧小区改造和宜居住区建设技术指南课题研究。

在课题研究的过程中和为各地老旧小区实践技术服务时我们注意到，当下的老旧小区改造存在一些普遍性的问题、困惑甚至误区，不仅是技术层面的，更有社会和机制层面的难点需要突破，也还需要在认识上进一步厘清，以及在理论上找到支撑。为此，我们在课题的基础上，结合江苏的多元实践探索以及应对当下社会经济发展和社区建设需求转变的思考，编著本书《江苏老旧小区改造建设导引》(简称《导引》)，目的在于揭示老旧小区改造的多维度内涵，引导老旧小区改造树立正确的价值导向，坚持以人民为中心的理念，通过共建共享共治，实现改善居民生活条件、促进城市结构优化、提升城市魅力的综合目标，让人民群众拥有实实在在的获得感和幸福感。

《导引》的主要特色和参考价值体现在五个方面：一是立足省情特点，通过实地调研和大数据分析，准确把握老旧小区的问题和居民需求，构建了系统综合、目标递进的改造内容体系，引导老旧小区改造因地制宜、精准施策；二是突显居民群众在老旧小区改造中的主体作用，通过“十大项、三层次”的分类分层指引，引导改造项目从以往的“政府排单”转向“群众点菜”，让居民主动参与更新改造；三是面向未来发展，积极融入绿色、健康、智慧、人文等多目标要求，鼓励老旧小区改造扩展视野，整合小区周边的存量资源，推动连片更新和区域整体发展；四是加强长效管理，将治理能力建设融入改造全过程；五是针对老旧小区改造参与主体的多元化特征，采用图文结合、便于理解的表达形式，使其成为向广大居民赋能的工具，让所有使用者都能获得启发和想象。

《导引》内容分为三部分：第一部分为“总体认识”，阐述老旧小区改造的背景意义、范围界定、江苏老旧小区的特点与需求、应当遵循的理念原则等。第二部分为“改造内容与建设导引”，包含“十大项、三层次”的改造菜单和技术指引，覆盖建筑、安全、基础设施、交通、环境卫生、便民服务、公共空间、绿化景观八项物质空间改造内容以及物业管理、长效机制两项制度建设内容；每项改造内容分为基础类、完善类、提升类三个层次，引导老旧小区结合自身情况，因地制宜进行差异化改造，多维度提升宜居品质。第三部分为“工作指引”，包括老旧小区改造的整体规划、一般流程、体检评估、共同缔造以及机制保障等内容，为科学有序地开展老旧小区改造工作提供参考。

《导引》既是一本面向老旧小区更新改造实施管理与规划设计相关从业者的技术指引，也是面向社会参与式更新实践的实用手册，我们希望每一位读者都能从中获得对老旧小区更新改造的认识和启发，并激发更多人对身边环境、城市发展进行关注和思考，对解决城镇化过程中的各种问题提出更多的、更有针对性的新思路、新理念。

本书虽是基于江苏实践的老旧小区改造建设指引，但其中的价值导向、内容体系、技术指引、机制建设对于全国其他地区的老旧小区改造也具有一定的借鉴和指导作用。我们希望能进一步加强与各方面同行的交流与合作，得到学术界与实践者们的批评与指正。

最后，特别感谢江苏省住房和城乡建设厅在课题编制过程给予的针对性指导，多次交流对于拓宽本书编写的思路提供了很大帮助！同时，课题编制在调研、意见征询等各个阶段得到了南京、扬州、盐城、苏州等地住房和城乡建设部门的积极配合和意见反馈，在此一并致谢！

江苏省规划设计集团总经理

2021年6月16日

CONTENTS
目录

6 方便居民日常生活
Make Daily Life More Convenient

7 以人为本改善公共活动空间
Improve Public Activity Space with People as the Core

10 建立长效机制 Establish Long-term Mechanism

PART 03 工作指引 / Work Guidelines

INSTRUCTIONS
使用说明

适用范围

本导引用于指导江苏省城镇老旧小区改造以及老旧小区周边环境提升；也可为其他省市城镇老旧小区改造及周边环境提升提供参考。

Scope of application

This guide is applicable to the renovation of old communities in cities and towns in Jiangsu and the upgrading of their surrounding areas; it can also serve as a reference for the renovation of old communities in cities and towns and the improvement of their surrounding areas in other provinces and cities.

使用对象

本导引主要面向组织实施老旧小区更新改造的相关部门、基层政府的管理人员以及参与老旧小区更新改造的设计机构、开发单位、社会组织、居民和个体。同时，本导引也是普通市民了解老旧小区改造、关注城市发展的普及类读物。

Users

This guide is primarily used by the management personnel in departments and primary level governments that organize and implement renewal and renovation of old communities, and the design institutes, social organizations, residents and individuals participating in the renewal and renovation of old communities. Meanwhile, this guide is also a popular reading material for ordinary citizens to know the renovation of old communities and take care of urban development.

使用方法

本导引由四部分组成，即使用说明、(第一篇)总体认识、(第二篇)改造内容与建设导引、(第三篇)工作指引。开展老旧小区改造工作，首先，依据第一篇理解老旧小区改造的背景、范围界定以及应当遵循的理念和原则；其次，参考第二篇所列的“十大项、三层次”的内容体系，结合小区自身情况，因地制宜合理生成改造项目，并采用适宜的技术方法实施改造项目；为科学有序推进老旧小区改造，可参考第三篇中的科学规划、工作流程、体检评估、共同缔造、保障机制等相关工作指引，保障老旧小区改造的有效实施。

Method of application

This guide consists of four parts, i.e. Instructions, (Part 1) General, (Part 2) Renovation Content and Construction Guide, and (Part 3) Work Guidelines. To carry out the renovation of old communities, firstly, according to Part 1, understand the background and scope of renovation of old communities, and the concepts and principles that should be followed; secondly, refer to the content system of “ten items with three levels” listed in Part 2, rationally generate renovation projects according to local conditions of communities, and adopt appropriate technical methods to implement the renovation projects; for the scientific and orderly advancement of the renovation of the old communities, please refer to scientific planning, work flow, inspection and evaluation, co-creation, and guarantee mechanism, etc. in Part 3 to ensure the effective implementation of the renovation of old communities.

各篇章主要内容

第1篇：

解读老旧小区改造的当代背景和意义，明晰老旧小区的范围界定，分析了江苏老旧小区的特点、问题、需求和未来价值，阐述了新背景下老旧小区改造的新导向、新理念以及工作目标、基本原则等内容。

第2篇：

构建了老旧小区改造的“十大项、三层次”内容体系，针对每项内容提出技术引导。

十大项：改善建筑质量、保障基础设施安全供应、消除安全隐患、改善交通及停车设施、保持小区环境整洁卫生、方便居民日常生活、改善公共活动空间、提升绿化环境景观、规范物业管理、建立长效机制。

三层次：每项改造内容分为基础类、完善类、提升类三个层次进行指引，其中基础类是指满足小区运行安全需要和居民基本生活需求的改造内容；完善类是指有条件的小区为满足居民生活便利和改善型生活需求，进一步完善功能、环境和管理的改造内容；提升类是指为丰富社会服务供给、提升居民生活品质，积极推进小区及周边空间整合利用，实现小区及周边环境整体优化的改造内容。

老旧小区改造应根据自身实际情况，因地制宜确定改造内容，有序递进，不断提升宜居品质，既要尽力而为，又要量力而行。

第3篇：

本篇是对老旧小区改造工作的指导，包括统筹编制老旧小区改造规划和年度计划，老旧小区改造的一般流程以及体检评估、共同缔造、保障机制等内容，为科学有序地开展老旧小区改造工作，促进居民参与，创新协同机制、实现美好家园共同缔造提供参考。

Main contents of parts

Part 1:

Understand the contemporary background and significance of the renovation of old communities, clarify the scope of the old communities, analyze the characteristics, problems, demands and future value of the old communities, and elaborate the new orientation, new concept, work goals and basic principles of the old community renovation under the new context. Ensure the effective implementation of the renovation of old communities.

Part 2:

A content system of “ten items with three levels” for the renovation of old communities has been constructed, and technical guidance is proposed for each item.

Ten items: improve construction quality, eliminate security risks, ensure the safe supply of infrastructure, improve traffic and parking facilities, keep the neighborhood clean and tidy, make daily life more convenient, improve the public activity space, enhance the environment landscape, standardize property management, and establish long-term mechanism.

Three levels: for each item, guide is made at the basic, improvement and upgrading levels, of which the basic aspect refers to those meeting the security needs for communities and the basic living demands of residents; the improvements refer to those to further complete the functions, environment and management in communities with ready conditions to facilitate the life of residents and improve their life; the upgrading aspect refers to those for integrated use of communities and surrounding space, realizing overall optimization of communities and surrounding environment to enrich the supply of social services, and raise the quality of life of residents.

In the renovation of old communities, the renovation contents should be determined according to the actual conditions and be made in an orderly and progressive manner, to continually raise the quality of life. Nevertheless, we should do our best within the capability.

Part 3:

Guide the renovation of old communities, including preparing the old community renovation planning and annual plan, the general process of the renovation of old communities, as well as the contents of inspection, evaluation, co-creation, and mechanism guarantees, etc., providing a reference to carry out the renovation of old communities in a scientific and orderly manner, promote residents' participation, innovate coordination mechanisms, and co-create a beautiful home.

1 背景意义 / Background and significance

2 范围界定 / Scope

3 特点与需求 / Characteristics and demands

4 转变与理念 / Change and philosophy

5 工作目标 / Objectives

6 基本原则 / Basic principle

PART 01 | 总体认识

General

1 背景意义
Background and significance

· 中共中央对“十四五”规划提出的建议明确提出实施城市更新行动，为今后一个时期做好城市工作指明了方向。老旧小区改造是城市更新行动的一种类型和有机组成部分，也是重大的民生工程和发展工程，2018年起，江苏连续三年将老旧小区综合整治列入省政府十大民生实事工程。2020年7月10日，国务院办公厅印发了《关于全面推进城镇老旧小区改造工作的指导意见》，要求全面推进城镇老旧小区改造工作，满足人民群众美好生活需要，推动惠民生扩内需，推进城市更新和开发建设方式转型，促进经济高质量发展。

· The suggestions made by the CPC Central Committee for the “14th Five-Year Plan” clearly put forward the implementation of urban renewal actions, and pointed out the direction for urban work in the future. The renovation of old communities is a type and organic part of the urban renewal action, as well as a major livelihood project and development project. Since 2018, Jiangsu has included the comprehensive renovation of old communities in the top ten livelihood projects of the provincial government for three consecutive years. On July 10, 2020, the General Office of the State Council printed and issued the *Guiding Opinions on Promoting the Renovation of Old Communities in Cities and Towns in All Aspects*, requiring that the work to renovate these old communities be pushed ahead in an all-round way, to meet the needs of the people for a better life, push the improvement of people' s wellbeing and expansion of domestic demand, advance the transition of city upgrading, development and construction methods, and promote the development of economy with high quality.

· 作为城市更新行动的重要组成部分，当下的老旧小区改造不仅要关注物质空间，还要关注社会层面，更要和城市发展相互促进，实施里子和面子、硬件和软件、内部和周边联动改造，探索住区、社区和城市互促之路。本《导引》即是基于上述需求进行研究和制定，旨在更有针对性地加强老旧小区改造工作的指导，提升老旧小区改造的实施效果。同时本导引也将在实践中不断完善和提升，让我们共同建设更安全、更方便、更舒心、更美好的生活家园。

· As an important part of the urban renewal action, the current old community renovation should not only pay attention to the physical space and the social level, but also promote each other with urban development, to implement the linkage renovation of inside and outside, hardware and software, internal and surrounding areas, and explore the mutual promotion of settlement, community and city. This guide has been developed on the basis of the above demands, for the purpose to enhance the guidance on the renovation of old communities in a targeted way and improve the implementation results of old community renovation. This guide also needs to be further completed and upgraded in practice. Let' s make joint efforts to make our living environment safer, more convenient, more comfortable and more beautiful.

2 范围界定
Scope

· 老旧小区是指城市或县城（城关镇）建成年代较早、失养失修失管、市政配套设施不完善、社区服务设施不健全、居民改造意愿强烈的住宅小区（含单栋住宅楼）。

· The old communities refer to residential communities (including a single residential building) that were built in a city or county seat (town) at an early age, with insufficient maintenance and management, incomplete municipal supporting facilities, imperfect community service facilities, and residents are strongly willing to renovate.

· 各地制定老旧小区更新改造计划，应综合考虑建造时间、客观情况、群众意愿、社会参与、财政能力等几个方面的要素。

· When formulating plans for the renewal and renovation of old communities, various localities should comprehensively consider factors such as construction time, objective conditions, popular will, social participation, and financial capabilities.

· **时间要求：** 重点改造2000年前建成的城镇老旧小区，2000年以后建成的小区也可纳入改造范围，但需限定年限和比例。

· **Time requirements:** focus on the renovation of old communities in towns and cities built before 2000, and those built after 2000 can also be included in the scope of renovation, but the number of years and proportions need to be limited.

· **客观情况：** 小区基础设施和功能明显不足、住区失养失修失管严重。

· **Objective conditions:** the community infrastructure and functions are obviously insufficient, and the residential area is seriously out of maintenance and management.

· **群众意愿：** 群众改造愿望强烈、参与程度高的优先纳入改造。

· **Popular will:** the priority is given to those with strong popular will and active participation in the renovation.

· **社会参与：** 对社会资本参与程度高的老旧小区，也要优先纳入改造。

· **Social participation:** the priority is given to those old communities with high social capital participation in the renovation.

· **财政能力：** 既要尽力而为，也要量力而行，不得盲目举债铺摊子。

· **Financial capacity:** We must do our best within our capability, rather than blindly take on debts.

3 特点与需求 Characteristics and demands

- 根据对江苏南、中、北不同地域代表性城市老旧小区现状及改造情况的实地调查研究，结合南京、宜兴、昆山等部分城市大数据调研分析，梳理江苏老旧小区的特点、问题、需求及其价值。总体上，江苏老旧小区由于历史演变，呈现面广量大、类型复杂、问题多样、需求差异的特征；另一方面，站在城市从外延走向内涵发展的高度，老旧小区的价值有待发掘，其将在很大程度上影响城市空间社会结构优化和发展竞争力。

- Based on field investigations on the status quo and renovation of old communities in representative cities in different regions of southern, central and northern Jiangsu, as well as big data research and analysis in some cities such as Nanjing, Yixing, Kunshan, etc., sort out the characteristics, problems, demands and value of old communities in Jiangsu. On the whole, due to historical evolution, the old communities in Jiangsu are characterized by a wide range, large quantity, complex types, diverse problems, and differences in demand; on the other hand, from the perspective of the city's development from the extension to the connotation, the value of the old communities remains to be explored, and they will affect the optimization of urban spatial social structure and development competitiveness to a large extent.

主要特点
Main characteristics

- **数量大、分布广**：江苏省2000年前建成的小区共有10411个，涉及居民302万户，主要分布于城市核心地段并享有便利的服务可达性，呈现出物质空间与社会环境“双重衰败”的显著特征，反过来也成为城市功能与社会治理“双重提升”的机遇空间。
- **类型复杂多样**：江苏老旧小区根据建设年代主要分为三类，第一类是建于新中国成立前后的街坊式住宅，多位于历史城区，具有一定保护价值；第二类是建于1960-1990年的各类新村、单元大院等住宅，既有成规模的小区，也有独栋零星的住宅，构成了老城区的基本肌理；第三类是建于1990-2000年城镇快速扩张阶段的拆迁安置房、保障房等政策类住房及小部分商品房。
- **空间资源有限**：第一类住宅以低层为主，多为传统院落式形态，产权关系复杂；第二类、第三类老旧小区一般为多层住宅，以4~6层为主，建筑密度普遍较高，用地形态亦多不规则，可以挖掘利用的存量资源非常有限，更新改造难度大。
- **重点研究对象**：第一类由于其历史价值，多会涉及历史文化保护范畴，在更新审批的流程上与其他老旧小区改造有所不同。本《导引》主要关注第二类、第三类具有典型性的老旧小区的改造研究。

- Large number and wide distribution: There were 10 411 communities built before 2000 in Jiangsu, involving 3.02 million residents. They are mainly distributed in the core areas of the city and enjoy convenient access to services, featuring obviously “double declines” of physical space and social environment, which, in turn, have become an opportunity space for the “double enhancement” of urban functions and social governance.
- Complex and diverse types: according to the construction age, the old communities in Jiangsu are divided into three types. The first is the neighborhood-style residential buildings built around 1949, mostly in historical urban areas, which are of certain protection value; The second is the new villages and unit courtyards built in 1960-1990, including both large-scale communities and single and sporadic residence buildings, which constitute the basic fabric of the old urban areas; The third is the policy housing, such as relocation and resettlement houses, and affordable houses, etc., and a small number of commercial housing built in the rapid expansion stage of cities and towns in 1990-2000.
- Limited space resources: the first type of residential buildings are mainly low-rise traditional courtyards involving complex property rights relationship; The second and third old residential areas are usually multi-storey houses with mainly 4-6 storeys, featuring in high building density and irregular land terrain, quite limited existing resources that can be excavated and utilized, and great difficulties in upgrading and renovation.
- Key research object: the first type is mainly related to the protection of history and culture because of its historical value. The process of renewal approval is different from the renovation of other old communities. This guide focuses on the research on the renovation of the second and third types of old communities with the typical characteristics.

3 特点与需求 Characteristics and demands

面临问题 Problems

- **物质环境差**：受制于早期的经济、技术、体制等因素影响，老旧小区当初建设标准较低，时至今日普遍存在房屋破损、基础设施老化、功能配套不足、公共空间缺乏等问题，难以满足居民日益增长的生活需求。
- **社会关系杂**：老旧小区多为工矿企业宿舍、行政事业单位房改房、拆迁安置房等类型，居民以退休老人、拆迁安置人口、外来人口等低收入弱势群体为主，素质良莠不齐，人口流动性高，容易引发邻里矛盾。
- **管理运营缺**：老旧小区大多没有物业管理和维修基金，安保措施缺乏，环境脏乱差，违章搭建、群租、无证开店等现象较为普遍，容易引发交通、消防、治安、卫生等一系列安全问题，亟待合理引导和整治。

- **Poor physical environment**: Due to the influence of early economic, technological, and institutional factors, etc., the construction standards of the old communities were lower then, and nowadays there are widespread problems such as house damage, aging infrastructure, insufficient functional facilities, lack of public space, etc., which make it difficult to meet the growing living needs of residents.
- **Miscellaneous social relations**: old communities are mostly dormitories for industrial and mining enterprises, houses of housing system transformation of entities and government-subsidized buildings for relocation settlement. Residents are mainly low-income vulnerable groups such as retired elderly, demolition and resettlement populations, and migrants, with varying quality and high population mobility and frequent conflicts in the neighborhood.
- **Lack of management and operation**: most of the old communities do not have property management and maintenance funds, with insufficient security measures, dirty, disorderly and bad environment, common existence of illegal construction, group rents, and unlicensed shops, which are prone to a series of security problems in traffic, firefighting, public security and sanitation, calling for the reasonable guide and renovation critically.

需求聚焦 Demand focus

- **住房性能改善**：居民首先关注住房安全，包括危房排查、房屋渗漏、楼道安全、违章搭建以及外立面构件安全等问题；其次，对于改善外观形象、加装电梯亦有较强的诉求。

- **Housing performance upgrading**: residents first take notice of housing safety, including dangerous house investigation, house leakage, corridor safety, illegal building and facade component safety; secondly, they have strong demands for improving the appearance image and adding elevators.

- **小区安全运行**：居民关注的顺序依次为水电气和路灯照明运行良好、公共环境定期保洁、公共空间使用安全、无高空抛物、无私拉乱接电线、消防通道畅通等，对于建立良好的安防管理比较关注。
- **停车设施规范**：老旧小区的人车冲突、停车难、停车无序问题是居民呼声最高、矛盾冲突最集中的一项内容，居民希望尽可能多的增加停车泊位、规范停车管理，让小区整洁有序，同时居民也希望尽量不要侵占绿化空间。
- **公共服务便捷**：居民希望在家门口就能得到多元化的公共服务，对于儿童、老人服务设施（幼儿园、社区卫生站、社区助餐点、社区活动中心等）以及便民商业（菜市场、药店、便利店、理发店、银行等），希望满足5分钟步行可达；对于康体运动设施，如健身器材、步道、球类场地、健身房，亦需求较高；同时对于公共厕所、快递柜等服务距离比较敏感。
- **公共空间品质**：居民普遍反对侵占公共绿地，希望改进绿化景观；对于户外活动场地，居民对配套设施较为关注，依次包括健身场地、儿童游戏场地、公共厕所、休息座椅、广场舞场地、无障碍设施等，希望能够在改造中予以完善；关注出行安全，对于周边的街道空间希望完善慢行通道、加强无障碍设施等人性化设计，让街道空间更加安全、舒适。
- **小区日常管理**：居民普遍希望加强信息沟通，拓宽信息沟通渠道，协商共议社区事宜；规范物业服务；发挥物业、居委会、街道在日常管理中的作用，加强社区应对各种灾害的能力。

- Safe operation of community: the residents' concerns are, in order, sound operation of water, electricity and street lighting, regular cleaning of public environment, safe use of public space, no highrise littering, no illegal wire connecting, and clear fire passage, etc. They pay more attention to the establishment of good security management.
- Specified parking facilities: the conflict between people and vehicles, parking difficulty and disordered parking in the old communities are the most popular appeal of residents and involve the most conflicts. Residents hope to add parking spaces as much as possible, standardize parking management, and make the residential area clean and orderly. At the meantime, residents also hope not to occupy the green space as much as possible.
- Convenient public services: residents hope to get access to diversified public services at home. They hope children and the elderly, service facilities (kindergartens, community health stations, community catering, community activity centers, etc.) and convenience businesses (vegetable markets, pharmacies, convenience stores, barbershops, banks, etc.) can be accessible within 5 minutes' walk. They also have a high demand for sports facilities such as fitness equipment, footpaths, ball game courts and gymnasiums. Meanwhile, they are sensitive to the service distance of public toilets and express cabinets.
- Quality of public space: residents are generally opposed to occupying public green space, but hope to improve the green landscape. For outdoor activity venues, residents pay more attention to supporting facilities, including fitness venues, children's playground, public toilets, rest seats, square dance venues, and barrier-free facilities, etc., and they hope these facilities can be improved in the innovation. They pay much attention to travel safety. In terms of the surrounding street space, they hope to improve the slow passage and strengthen the humanized design of barrier free facilities, so as to make the street space safer and more comfortable.
- Community daily management: residents generally hope to strengthen information communication, broaden information communication channels, and discuss community matters through consultation; standardize property services; give full play to the role of property management company, neighborhood committees and street office in daily management, and strengthen the community's ability to cope with various disasters.

3 特点与需求 Characteristics and demands

未来价值 Future value

- **降低城市化成本**：老旧小区在我国城市化过程中发挥了重要作用，对其更新改造可有效缓解整体拆迁安置成本高的实际困难，有利于稳定房价，对于促进外来人口市民化具有积极意义，也可有效避免城市发展中的贫民窟和绅士化问题，促进社会公平。
- **保护历史记忆和文脉**：老旧小区改造可避免大拆大建造成城市肌理的断裂，通过挖掘低效闲置资源进行修复修补以及再开发等微更新手段，可以不断丰富城市功能和形态，提升和再造城市魅力，为城市发展注入更多的活力。
- **促进社区经济发展**：老旧小区改造可在养老、托育、医疗和家政服务等行业创造巨大的发展空间，创造更多的就近就业空间，提供类型丰富且便捷可达的社区服务，有效带动社区产业、社区文化、社区服务等多个层面的全面提升。
- **培育基层治理能力**：老旧小区因为建成时间长，居民邻里之间相熟，形成了相对稳定的社会关系和社区生活模式，在推动社区治理能力建设的过程中有利于结成共同体，培育社区自我发展和自我更新能力。

- **Reduce the cost of urbanization**: Old communities have played an important role in the process of urbanization in China, and their renovation can effectively alleviate the actual difficulties of high costs of overall demolition and resettlement, help stabilize housing prices, play a significant role in promoting the citizenization of immigrant population, and avoid slums and gentrification in urban development and promote social equity.
- **Protect historical memory and context**: The renovation of old communities can prevent large-scale demolition and large-scale construction from breaking the urban fabric. By mining inefficient idle resources, repairing, and redevelopment and other micro-renewal methods, the urban functions and forms can be continuously enriched, and urban harmony can be improved and re-created, thus injecting more vitality into urban development.
- **Promote the economic development of the community**: the renovation of old communities can create huge development space in industries such as elderly care, nursery care, medical care and housekeeping services, create more nearby employment spaces, provide a variety of convenient and accessible community services, and effectively drive the comprehensive improvement in multiple levels of community industry, community culture, and community services.
- **Cultivate primary level governance capabilities**: As the old communities have been built for a long time, residents and neighbors have become familiar with each other, forming a relatively stable social relationship and community life mode, which, in the process of promoting community governance capacity building, is conducive to forming a community and nurturing community self-development and self-renewal ability.

4 转变与理念
Change and philosophy

导向的转变
Change in orientation

· 党中央和国家长期以来坚持以人民为中心的发展思想，高度重视人居环境改善和人民群众生活品质提升。近年来从棚户区改造到城中村和危房改造，再到城镇老旧小区改造，出台了一系列的国家政策。2019年以来，国务院常务会议、政治局会议等多次重大会议均提出要大力进行老旧小区改造提升。2020年7月，国务院办公厅颁布《关于全面推进城镇老旧小区改造工作的指导意见》，明确了城镇老旧小区改造是重大民生工程和发展工程，老旧小区改造成为新时代城市更新工作的使命和任务。

· 我国城镇老旧小区量大面广，涉及上亿居民，不可能一蹴而就，需要久久为功。作为城市更新的一种类型和有机组成部分，老旧小区改造涉及城市物质环境以及社会、经济诸多方面，既是一项专业性强的技术工作，也是一项政策性强的社会工作。老旧小区改造不仅要关注物质空间，还要关注社会层面，更要立足城市视角，与所在地区乃至城市发展战略进行关联思考，探索住区、社区和城市互促之路，根本目的是为居民营造一个和谐宜居、方便高效、健康卫生、优美且富有文化内涵的人居环境。

· The CPC Central Committee and the State Council have always adhered to the people-centered development concept, and attached great importance to the improvement of the living environment and the quality of life of the people. In recent years, a series of national policies have been promulgated, involving from the renovation of shanty towns to the renovation of urban villages and dilapidated houses, and to the renovation of old communities in cities and towns. Since 2019, many major meetings, such as the executive meeting of the State Council and the Political Bureau meeting, have proposed to vigorously renovate and upgrade old communities. In July 2020, the General Office of the State Council promulgated the *Guiding Opinions on Promoting the Renovation of Old Communities in Cities and Towns in All Aspects*, which clarified that the renovation of old communities in cities and towns is a major livelihood project and development project, and the renovation of old communities has become the mission and task of urban renewal in the new era.

· The old communities in cities and towns involve wide aspects and hundreds of millions of residents in China. The renovation cannot be accomplished overnight but take a long time. As a type and organic part of urban renewal, the renovation of old communities involves many aspects such as the urban material environment, society and economy. It is not only a highly professional technical work, but also a policy-related social work. The renovation of old communities should pay attention not only to the physical space, but also to the social level, correlate to the development strategy of the region and even the city from the urban perspective, and explore the way of mutual promotion among settlement, community and city. The fundamental purpose is to create a harmonious, livable, convenient, efficient, healthy, beautiful, and cultural living environment for residents.

4 转变与理念 Change and philosophy

- **从短期工程转向有机更新**：传统大拆大建、立面整治、环境美化等运动式改造模式，由于缺乏维护和长效性，时间一长难免再次衰败；未来应当融入居民主体，通过有机修补和开发利用，实现从空间、功能等物质层面到人文、社会非物质层面的系统更新，注重完整可持续社区营造，促进区域活力提升。
- **从局部思维转向系统改造**：以往头痛医头、脚痛医脚的碎片化、局部化改造方式，改造效果有限，居民获得感不强；未来应当注重功能的系统性更新和宜居性的整体性提升，尊重和营造社区文化，让居民拥有实实在在的获得感和幸福感。
- **从单个小区转向层级联动**：现阶段老旧小区改造大多局限于项目本身，较少与所在区域乃至城市发展战略进行关联思考，难以有效发挥以点带面促进地区整体发展复兴的作用。未来需要拓宽改造视野，探索小区－街区－社区－城市多层级互动，推动连片更新，最终实现城市整体宜居水平的提高。
- **从政府主导转向共同缔造**：老旧小区改造涉及多元主体，单一的政府主导不具有可持续性。老旧小区改造应当重视社区营造，积极探索政府引导，居民、市场、社会多方参与，民主决策的更新机制，构建多样化的协同平台，引导居民从被动接受改造到主动参与更新，推动社区的共建共享共治，完善长效治理机制。

- **Change from short-term projects to organic renewal:** in the traditional campaign-type renovation modes, such as large-scale demolition and large-scale construction, facade renovation, and environmental beautification, etc., the communities will inevitably decline again over time due to the lack of maintenance and long residual action. In the future, the renovation should be integrated into the residents entities, realize the system renewal from the material level of space and function to the non-material level of humanity and society through organic repair, development and utilization, attach importance to the construction of a complete sustainable community, and enhance the regional vitality.
- **Change from local thinking to system improvement:** In the past, the fragmented and localized renovation methods led to the limited renovation effect and the insufficient sense of gain of residents. In the future, attention should be paid to the systematic renewal of functions and the integrated improvement of livability. Respect and create a community culture to render residents with a real sense of gain and happiness.
- **Change from a single community to hierarchical linkage**: At the current stage, the renovation of old communities is mostly confined to the project itself, but seldom correlates to the development strategy of the regional and the city, so it is difficult to effectively play the role of promoting the overall development and revival of the region. In the future, we shall broaden the vision of renovation, explore the multi-level interaction of neighborhood, block, community and city, and promote contiguous renewal, and finally improve the overall livability of the city.
- **Change from government leading to co-creation:** The renovation of old communities involves multiple subjects, and a single government leading is not sustainable. The renovation of old communities should build up the community, actively explore the renewal mechanism of government guidance, participation by residents, the market and the social forces, and democratic decision-making, build a diversified collaborative platform, guide the residents to change from passive acceptance of renovation to active participation in renewal, and promote community co-construction, sharing and co-governance, and enhance long-term governance mechanisms.

与时俱进的理念
Concept of advancing with the times

- **民生为本，满足多样化生活需求**：坚持问题导向、需求导向，因地制宜精准施策；面向未来发展，融入健康、绿色、智慧、人文等多目标要求，以微更新、渐进式的策略应对更加多样化的生活需求；挖掘和尊重地方特色，营造多样化、在地化的社区景观和文化认同，增进社区归属感和自豪感。
- **系统思维，提升综合效益**：老旧小区改造应与城市更新战略相结合，统筹协调好硬件和软件、地上和地下、近期和长远、内部和周边的关系，集成联动实施改造，保证改造的综合效益，不断探索更新技术，推动住区更新产业发展。
- **城市视角，推动连片更新**：老旧小区不是独立封闭的空间范畴，需立足周边及城市统筹发展的角度，挖掘存量资源，修补城市所缺，建设完整社区，通过连片更新、层级联动，不断优化社会空间结构，促进区域活力提升。
- **融入治理，促进社区复兴**：老旧小区改造过程也是城市治理能力提升的过程，需要建立运营思维，推动居民参与、共同缔造，完善上下结合的长效管理机制，才能让老旧小区改造提升长期持续地推动下去。

- **Oriented to the people's wellbeing, and meeting diversified living needs:** The renovation of old communities shall be problem-oriented, demand-oriented, and carry out precise policies based on local conditions, develop towards the future, integrate multiple goals such as health, green, wisdom, and humanities, and respond to more diversified living demands with micro-update and gradual strategies; explore and respect local characteristics, create a diverse and localized community landscape and cultural identity, and enhance the sense of belonging and pride in the community.
- **Systematic thinking, improving comprehensive benefits:** The renovation of old communities should be integrated with the urban renewal strategy, and the correlations between hardware and software, overground and underground, short-term and long-term, internal and surrounding areas should be coordinated, implement the renovation in an integrated and linked manner to ensure the comprehensive benefits of the renovation, continuously explore and update technology, and promote the renewal of residential area and the industrial development.
- **Urban perspective, promoting contiguous renewal:** The old communities are not an independent and enclosed space category. The renovation of old communities shall, based on the perspective of the surrounding and urban overall development, tap the existing resources, mend what the city is lacking, and build a complete community. Through contiguous area renewal and hierarchical linkage, optimize the social space structure and enhance the regional vitality.
- **Integrate into governance and promote the community revitalization:** The renovation of old communities is also a process of improving urban governance capabilities. We should establish operational thinking, promote residents' participation and co-creation, and improve the long-term management mechanism combining upper and lower levels, so that the renewal and upgrading of old communities can be pushed forward for a long time.

5 工作目标
Objectives

- **改善住区生活环境：** 提升老旧小区建筑质量、基础设施、公共服务配套和环境品质，营造安全、健康、便利、人性化的生活空间。
- **带动周边整体提升：** 整合老旧小区周边资源，加强内外空间链接，与宜居街区建设相衔接，促进地区整体品质提升。
- **健全物业服务机制：** 建立托底保障机制，引导物业服务市场整合和创新发展，提升物业行业整体服务能力。
- **完善社区治理体系：** 加强公众参与，践行共同缔造，推动社区治理水平提升。

- **Improve the living environment of residential areas:** improve the building quality, infrastructure, public service facilities and environmental quality of old communities, and create a safe, healthy, convenient and humanized living space.
- **Promote the overall upgrading of surrounding areas:** integrate the surrounding resources of old communities, strengthen the internal and external space link, connect with the construction of livable blocks, and promote the overall quality upgrading of the region.
- **Improve the property service mechanism:** establish a basic support mechanism, guide the integration and innovative development of the property service market, and enhance the overall service capability of the property industry.
- **Improve the community governance system:** strengthen public participation, practice joint creation, and promote the improvement of community governance.

6 基本原则 Basic principle

· **问题导向、民生优先：** 坚持问题导向、需求导向，针对老旧小区现状进行体检评估，以群众居住生活领域急需解决的问题为出发点，切实改善居民生活质量，营造安全、健康、全龄友好、和谐文明的宜居住区。

· **因地制宜、差异改造：** 结合老旧小区的实际情况、资源条件以及问题的迫切程度，实事求是合理确定改造内容、改造时序，从以往建设项目的“政府排单”转向“群众点菜”，突出针对性，力求使改造成果获得居民认同。

· **节约集约、普惠百姓：** 老旧小区改造需结合各地发展实际，既要尽力而为，又要量力而行，优先解决安全、养老、公共服务、物业管理等最为迫切的问题，逐步提升居住品质，避免公共财政资金浪费。

· **多方参与、共同缔造：** 积极探索政府推动、市场参与、人民主体的改造模式，激发居民参与改造的积极性、主动性，充分调动社会力量支持和参与老旧小区改造；突出基层党建的引领示范作用，将社区治理能力建设融入改造过程，完善小区长效管理机制。

· **Oriented by problems and take people's wellbeing as priority:** We should adhere to orientation by problems and demand, inspect and evaluate the status quo of old communities, definitely improve the quality of life of residents with solving the urgent needs of masses in living areas as the start point, to create safe, healthy, harmonic and culturally developed livable residential areas friendly to people at all ages.

· **Act according to local conditions and carry out the differentiated renovation:** The contents and sequence of renovation shall be determined reasonably according to the actual conditions and resources of old communities and the urgency of problems, shifting from the previous “government arrangement” to “ordering by masses” for construction projects, to ensure targeted renovation and that the renovation results are accepted by residents.

· **Economical and intensive to inclusively benefit the people:** The renovation of old communities should be based on the actual development of local area with the best efforts and within the capacity, and priority shall be given to most urgent problems in security, old-age caring, public service and property management, to gradually improve the residence quality and avoid waste of public funds.

· **Joint work with multi-party participation:** Active exploration shall be made for the renovation pattern of government pushing and market participation with people as the main body, to activate the initiative of residents to participate in renovation, and fully mobilize social forces to support and participate in the renovation; the leading and demonstration role of primary level Party organization shall be highlighted, to merge the construction of community governance capability into the renovation and complete the long-term management mechanism of community.

1 改善建筑质量 / Improve Construction Quality

2 消除安全隐患 / Eliminate Security Risks

3 保障基础设施安全供应 / Ensure the Safe Supply of Infrastructure

4 改善交通及停车设施 / Improve Traffic and Parking Facilities

5 保持小区环境整洁卫生 / Keep the Neighborhood Clean and Tidy

6 方便居民日常生活 / Make Daily Life More Convenient

7 以人为本改善公共活动空间 / Improve Public Activity Space with People as the Core

8 提升绿化环境景观 / Enhance the Environment Landscape

9 规范物业管理 / Standardize Property Management

10 建立长效机制 / Establish Long-term Mechanism

PART 02 | 改造内容与建设导引

Renovation Content and Construction Guide

Construction Guidance for Renovation of Old Communities in Jiangsu

江苏老旧小区改造建设导引

1

改善建筑质量
Improve Construction Quality

- 1.1 保证房屋正常安全使用
  Ensure the normal and safe use of houses
- 1.2 整治违章搭建
  Remedy illegal setups
- 1.3 改善建筑外观形象
  Improve building appearance
- 1.4 提高节能性能
  Raise energy saving performance
- 1.5 鼓励加装电梯
  Encourage adding elevators

1.1 保证房屋正常安全使用
Ensure the normal and safe use of houses

房屋应进行危险性排查
Check houses to find out danger

- 街道社区应定期摸排小区房屋质量情况，对于涉危房屋，应委托具有相应资质的专业机构，根据相关标准，进行危险性等级鉴定。
- 针对危房不同情况，制定相应改造方案，确保小区内不存在C、D级的危房。
- 对于可能涉及蚁害的房屋，应由白蚁防治单位对房屋的蚁害情况进行调查，采取相应措施进行灭治与预防。

- Communities shall periodically check the quality of houses, for those with possible danger, professional institutions with corresponding qualification shall be entrusted to determine the class of danger according to relevant standards.
- For houses with different danger, corresponding renovation plan shall be worked out, to ensure that there is no house with danger of classes C and D in the community.
- For houses that may be damaged by termite, have a termite control unit investigate the situation of damage of the houses and take corresponding measures to kill termite and prevent the damage.

外墙斜裂缝

内墙贯穿裂缝

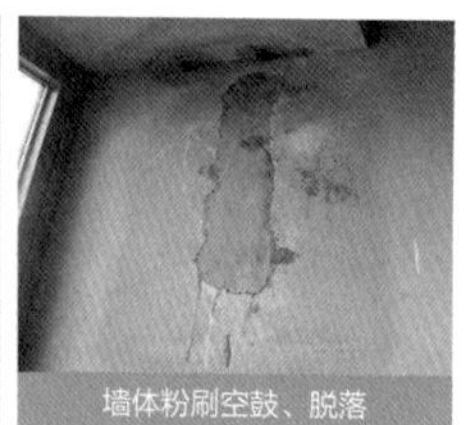
墙体粉刷空鼓、脱落

楼体局部塌陷

窗台下方墙体斜裂缝

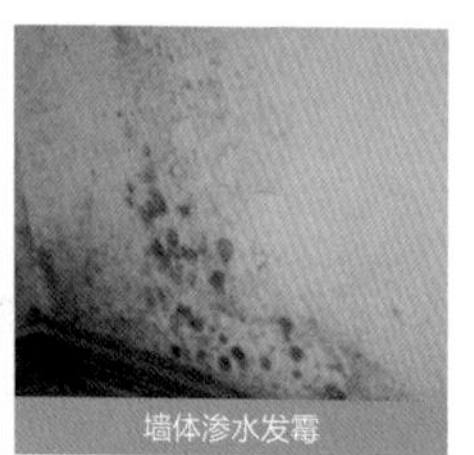
墙体渗水发霉

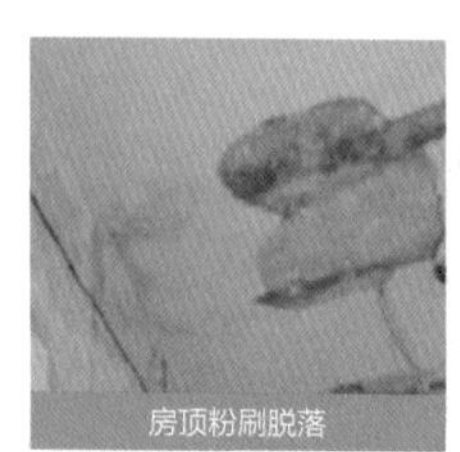
房顶粉刷脱落

屋顶瓦片滑移

混凝土过梁开裂

图2-1-1 常见的房屋安全隐患 / Common safety hidden dangers in houses
| 图片来源：编写组自摄

- 根据住房和城乡建设部提出的危险房屋鉴定标准：
  A级：结构承载力能满足正常使用要求，无腐朽危险点，房屋结构安全。
  B级：结构承载力基本满足正常使用要求，个别结构构件处于危险状态，但不影响主体结构，基本满足正常使用要求。
  C级：部分承重结构承载力不能满足正常使用要求，局部出现险情，构成局部危房。
  D级：承重结构承载力已不能满足正常使用要求，房屋整体出现险情，构成整幢危房。

- Standards for determining houses with danger as put forth by the Ministry of Housing and Urban and Rural Construction:
  Class A: the structure bearing capacity can meet the requirements of normal use, free of decay and dangerous point, and the house structure is safe.
  Class B: the structure bearing capacity basically meet the requirements of normal use, a small number of structural members are in danger, but not affecting the main structure, basically meeting the requirements of normal use.
  Class C: the bearing capacity of some load-bearing structure cannot meet the requirements of normal use, with local dangerous condition, as house with local danger.
  Class D: the bearing capacity of load-bearing structure cannot meet the requirements of normal use, with dangerous condition in the whole house, rendering the whole house in danger.

1.1 保证房屋正常安全使用
Ensure the normal and safe use of houses

图2-1-2 局部加强保护 / Local reinforcement | 图片来源：编写组自摄

图2-1-3 整体重做防水 / Overall replacement of water-proof layer | 图片来源：编写组自摄

图2-1-4 平改坡的处理方式 / Change flat roof to sloped roof | 图片来源：编写组自摄

图2-1-5 坡屋面与立墙交界部位加强保护 / Enhanced protection at the interface of sloped roof and vertical wall | 图片来源：编写组自摄

图2-1-6 坡屋面更换可靠的新型防水材料 / Replace reliable new type water-proof material for sloped roof | 图片来源：编写组自摄

▶ 改善房屋渗漏
Stop leakage for houses

· **屋面防水处理：**针对平屋面或者坡屋面，采用不同的方式进行防水处理。

平屋面防水处理：
针对屋面渗漏部位进行局部改善加强保护；
屋面防水材料老化情况严重的，宜整体重做防水；
位于重点地段、在符合房屋主体结构安全前提下，可实施平改坡的处理方式。根据原有房屋的结构承载力情况，增建的坡屋顶主体结构宜采用轻钢龙骨架等自重较轻的结构材料。

· **Water-proof treatment for roofing:** water-proof treatment shall be made in different ways for flat or sloped roofs.

Water-proof treatment for flat roofs:
Partial improvement for more protection for the leaking part on roof;
If the roof water-proof material has seriously aged, it should be fully replaced;
Flat roof can be changed into sloped type for those in key areas while ensuring the safety of the main structure of the house. According to the bearing capacity of the original house structure, the main structure of the added sloped roof should be made of materials in light weight such as lightgate steel joist.

坡屋面防水处理：

针对屋面卷材损坏部位进行局部修补；

加强坡屋面与立墙交界部位的防水保护；

更换坡屋面防水材料，鼓励采用防水性能好、耐久时间长的新型防水材料，对坡屋面进行防水修缮。

Water-proof treatment for sloped roofs:

Local repair of the damaged sheet on roof:

Water-proof protection at the interface of sloped roof and vertical wall should be reinforced;

Roof water-proof material should be replaced, and it is encouraged to repair the sloped roof with new type water-proof materials with good performance and long durability.

- **外窗渗漏处理：** 先清理老化材料，后重做防水。
- **外墙渗漏处理：** 先修补墙面基层，再进行防水处理。

- **Leak treatment for external windows:** first remove the aged material, and then make water-proof layer.
- **Leak treatment for exterior wall:** first repair the wall base, and then make water-proof treatment.

图2-1-7 墙面改造过程及效果 / Wall renovation and effect | 图片来源：编写组自摄

图2-1-8 外窗更换处理 / External window replacement | 图片来源：编写组自摄

1.1 保证房屋正常安全使用
Ensure the normal and safe use of houses

保障房屋公共部位整洁、安全、有序
Ensure the public part of houses clean, safe and in good order

- 楼梯间出新改造，达到整洁、明亮、通畅的要求。
  清理楼梯走道乱堆杂物；
  清理墙面乱贴的小广告；
  修补粉刷老旧破损的墙面、顶棚和踏面；
  完善楼梯间照明设施；
  整理归并楼梯间的管线及各类生活设施（如：信箱、奶箱等）；
  因地制宜增加墙面美化内容。
- Renovate the staircase, to make it clean, tidy, bright and well-through.
  Clean off sundries in staircase and walkway;
  Remove small advertisements pasted on wall;
  Repair and whitewash old and damaged wall, ceiling and tread;
  Complete the lighting facilities in staircase;
  Re-arrange and put together pipelines and various living facilities (such as mail-boxes and milk boxes) in the staircase;
  Beautify the walls according to actual conditions.

图2-1-9 老旧小区楼梯间常见问题 / Common problems in staircases of old communities | 图片来源：编写组自摄

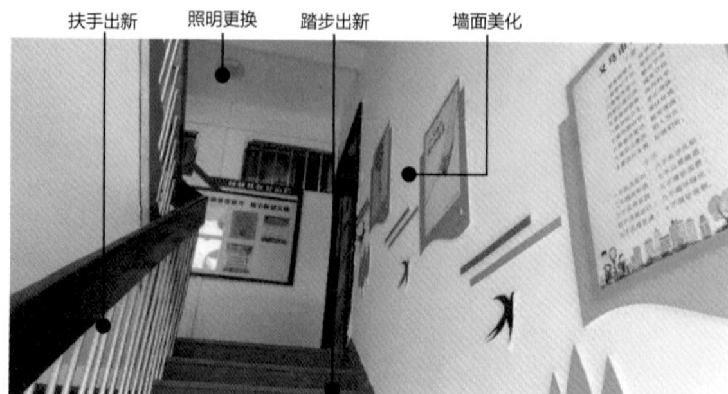

图2-1-10 楼梯间出新改造措施 / Repair and renovation measures for staircase | 图片来源：编写组自摄

· 楼梯间设施改造，杜绝安全事故。
修补出新老旧破损的栏杆及扶手；
首级踏步增加醒目标识，提升安全性；
有条件的地区，可增加便于老人使用的方便设施，包括增设双侧扶手、休息平台增设方便座椅、首层踏步的坡道化处理等。

· Renovate facilities in staircase to prevent safety accident.
Repair and replace old and damaged railing and handrails;
Add conspicuous signs on the first step for more safety;
Where conditions permit, convenient facilities for use by the elderly can be added, including handrails on both sides, chair at rest platform and sloped step on the first floor.

图2-1-11 楼梯间安全改造措施 / Safety renovation measures for staircase | 图片来源：编写组自摄

· 上人屋面的栏杆需核查防护高度，从可踏面起计算，应大于1.1m，栏杆的杆件间距应满足规范要求。

· The protection height of railing for manned roof should be inspected, it shall be over 1.1m from the accessible step surface, and the spacing between railing elements shall meet the specification.

图2-1-12 上人屋面栏杆防护高度应从可踏面算起 / The protection height of railing for manned roof shall be counted from the accessible step surface | 图片来源：编写组自摄

1.2 整治违章搭建
Remedy illegal setups

违章搭建的普查
General survey of illegal setups

· 老旧小区实施改造前，街道社区联合相关部门组织开展小区内违章搭建物的普查工作，编写违建手册。

· 违章搭建的范围包括：未取得建设工程规划许可证建造的建筑物、构筑物；未按照建设工程规划许可证核准的图纸及相关技术规定建造的建筑物；以及擅自在天井、庭院、平台、晒台（露台）、屋顶、道路或其他场地建造的建筑物、构筑物，开挖地坪以及在房屋内部插层增加的建筑面积、擅自改变建筑物原规划设计外观等情况。

· Before the renovation of an old community, the community shall organize a general survey of illegal setups in the community together with relevant departments, and prepare a list of illegal setups.

· Illegal setups include: buildings and structures constructed without obtaining the construction project planning permit; buildings constructed not according to the drawings and relevant technical specifications approved by the project planning permit; and buildings and structures constructed in courtyards, platforms, flat roofs (balcony), roads or other ground without permission, building areas added by excavating ground and adding floor inside the house, or by changing the original design appearance of the buildings.

违章搭建的确认
Confirmation of illegal setups

· 街道社区与建设单位共同核对违建清册，同时对小区周边、影响小区整治效果却未被列入清册的违章建筑，应与街道确认后纳入违建清册。

· The community and owner entity shall jointly check the list of illegal setups, and the illegal setups around the community and affecting the rectifying effect of the community but not on the list shall be included in the list of illegal setups after confirmation with the community.

违章搭建的整治原则
Principles in demolishing illegal setups

· 坚持“应拆尽拆”的原则，保证老旧小区改造的整体效果。

· 对于因特殊历史原因建成的建、构筑物，在保证其安全性的前提下，可与主管部门协商，通过特殊程序解决。

· Adhere to the principle of “demolishing all that should be demolished”, to ensure the overall effect of old community renovation.

· For buildings and structures constructed due to special causes in history, solution can be made through consultation with department in charge according to special procedures while ensuring the safety.

违章搭建的清理实施
Demolish illegal setups

- 相关责任部门、街道社区、建设单位共同制定拆违方案和计划，同时明确责任人和截止时间。
- 街道社区牵头组织召开小区拆违工作居民议事会，公示违建清册内容，同时做好违建拆除的宣传发动、矛盾协调、维稳等工作。
- 相关责任部门牵头，与街道社区、城管、公安、消防等部门联合执法，对违建进行拆除。拆除过程应做好影像资料留存。
- 拆违期间，责任人应做好拆违垃圾的集中堆放、及时覆盖、扬尘控制等工作，拆违垃圾必须日产日清，不影响居民出行和生活。

- The relevant responsible departments, community and owner entity shall jointly work out plan and scheme to demolish illegal setups, and also specify responsible person and time limit.
- The community shall take the lead to organize a meeting of residents in the community on demolishing illegal setups, publicize the contents of the list of illegal setups, and also properly do work in publicity, conflict coordination and stabilizing people in the demolishing.
- The relevant responsible departments shall take the lead to jointly enforce law with the community, city administration, public security and fire protection to demolish the illegal setups. Photo files shall be reserved for the demolishing process.
- During demolishing, responsible persons shall properly arrange work in collecting wastes, timely coverage and dust control, wastes from demolishing must be removed on the current day, without affecting the traffic and life of residents.

违章搭建的长效管理
Long-term management over illegal setups

- 街道、社区、物业作为违建长效管理的责任主体，应加强对小区整治后的巡视管理，避免违建现象的再次发生。

- As the main bodies for long-term management of illegal setups, the subdistrict, community and property management shall strengthen the tour inspection of the community after the rectification, to avoid recurrence of illegal setups.

案例参考 Reference cases

坚决拆除：针对违法占用消防车道、危害公共安全、影响住宅日照的违章搭建，依据法定程序坚决予以拆除。
Firmly demolish: Those illegally taking firefighting lanes, endangering public safety and affecting sunshine of residences shall be firmly demolished according to legal procedures.

暂时保留：针对困难家庭底层院落、顶层露台擅自加盖屋顶的违章搭建现象，暂时保留，协助其对构件进行安全加固、外观美化。参考小区“扬州荷花池社区”“南京马鞍山3号院”“南京天顺苑”。
Retain tentatively: Illegal setups such as those in courtyard or roof balcony added with roof in households with difficulties are retained tentatively, and they were assisted to reinforce the structure and beautify the appearance. Reference communities: “Hehuachi Community in Yangzhou” “Maanshan No. 13 Courtyard in Nanjing” and “Tianshun Garden in Nanjing”.

转换为公共设施：针对小区房屋前存在的私自搭建自行车棚现象，予以统一改造，转换为方便居民使用的公共自行车棚。参考小区“南京宁夏路18号”“南京宁夏路13号”。
Conversion into public facilities: Bicycle sheds built without permission in front of buildings were transformed on a unified basis into public bicycle sheds for the convenience of residents. Reference communities: “18 Ningxia Road in Nanjing” and “13 Ningxia Road in Nanjing”.

图2-1-13 案例参考示意图 / Case reference diagram | 图片来源：编写组自摄

1.3 改善建筑外观形象
Improve building appearance

图2-1-14 通过平改坡的方式对屋面进行加固美化 / Reinforce and beautify roof by changing flat roof into sloped one | 图片来源：编写组自摄

图2-1-15 扬州康乐社区外墙局部脏污出新 / Local renovation of dirty external walls in Kangle Community in Yangzhou | 图片来源：编写组自摄

图2-1-16 盐城东台东坝新村外墙线缆序化 / Rearrange wires and cables on external walls in Dongba New Village, Dongtai of Yancheng | 图片来源：编写组自摄

▶ 保持建筑外观干净整洁
Keep the building appearance clean and tidy

· 建筑外观修缮应遵循安全、美观、节能、环保、符合区域风貌控制规划的原则。

· Renovation of building appearance shall follow the principle of safety, good look, energy saving, environment and conforming to the regional style control planning.

· 屋面加固美化处理：进行屋面防水处理后，整理序化屋面设施，更新防雷设施，保证屋面整洁美观。有需要时适当进行局部加固。

· Roof reinforcing and beautifying: after water-proof treatment of roof, the roof facilities shall be rearranged and lightning protection facilities replaced, to ensure cleanness and good look of roof. When necessary, local reinforcement will be made.

· 外观破损脏污处理：根据墙面脏污破损的情况，采用局部修补或整体粉刷的方式进行出新。

· Treatment of damaged and dirty appearance: local repair or overall whitewashing shall be made according to the damage or dirt on walls.

· 外墙线缆序化处理：外墙上的线缆应统一整理，可采用立管将电力、通信等线缆包围隐蔽，做到齐整不杂乱，保证建筑外观洁净有序。

· Wires and cables on external walls shall be re-arranged on a unified basis, and the power and communication wires can be concealed with vertical pipes, so that they are in good order to ensure a clean and orderly appearance of buildings.

▶ 外立面构件应安全整齐美观
Structural members on facade shall be safe, tidy and beautiful

- **空调机位的处理**：
  通过移位使空调外机摆放整齐有序；
  设置统一的空调机罩构架使其和谐美观；
  结合建筑立面整体设计，提升建筑整体形象；
  空调机罩的装饰栏杆宜选用透空率良好的百叶或者金属杆件，不影响空调外机的正常散热性能。

- Arrangement of air conditioners:
  Relocate the air conditioner external units to arrange them in good order;
  Provide unified air conditioner hood and racks to ensure harmony and good appearance;
  Improve the overall image of the building in conjunction with the overall facade design for the building;
  Shutters with good permeability or metal rods should be used as railing for air conditioners, so that normal heat dissipation of air conditioner external units is not affected.

图2-1-17 统一设置空调机罩构架 / Provide air conditioner hood and racks on a unified basis | 图片来源：编写组自摄

图2-1-18 结合建筑立面整体设计空调机位 / Design the air conditioner locations in conjunction with the overall facade design of buildings | 图片来源：编写组自摄

1.3 改善建筑外观形象
Improve building appearance

外立面构件应安全整齐美观
Structural members on facade shall be safe, tidy and beautiful

· 防盗网、晾晒架、遮阳棚、窗台花架、阳台栏杆等构件的处理：
应结合建筑立面统一设计，体现整体性；
所用的外立面构件材质应符合安全标准；
防盗网需预留人员安全出口，并与房屋整体风格相协调。

· Treatment of security net, air drying racks, awning, windowsill flower racks, balcony railing, etc.
Unified design with the building facade, to embody an overall effect;
The materials for all external facade members shall comply with safety standards;
Security net shall be provided with exit for personnel, and be harmonic with the overall style of the house.

图2-1-19 遮阳棚（需牢固、美观）/ Awning (firm and in good appearance) | 图片来源：编写组自摄

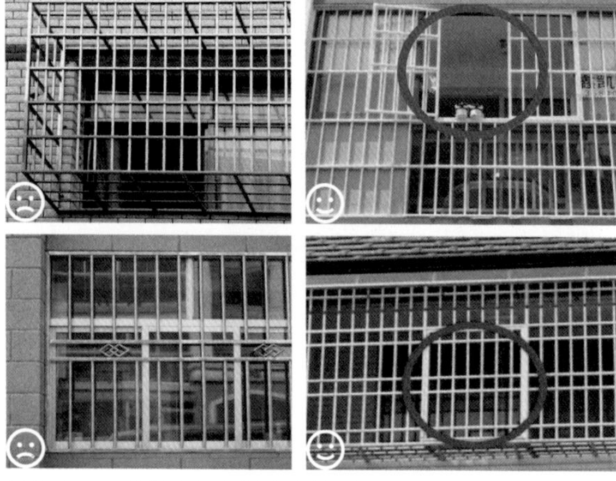

图2-1-20 防盗网（需带安全出口）/ Security net (security exit shall be provided) | 图片来源：编写组自摄

图2-1-21 晾晒架（需统一、牢固）/ Air drying racks (in unified and firm structure) | 图片来源：编写组自摄

底层院落整治

Renovation of courtyards

- 统一出新优化院墙，院墙高度应综合考虑视线、景观等要求，一般宜为1.6~2.0m。
- 院墙上部应适当考虑镂空设计，防止违章搭建。
- 院墙形式、色彩、材质应与房屋、周边环境相协调，鼓励因地制宜进行院墙美化，彰显小区文化特色。
- 有条件的可以统一增设防坠落顶棚。
- 底层院落入口处可增加入口雨篷，强化标识。
- 通过增设绿化景观、装饰细节，美化入口空间。

- Build new courtyard walls on a unified basis, the wall height should be 1.6~2.0m in general by taking into overall account of sight and landscape.
- Hollow-out design shall be used as appropriate for the upper part of courtyard wall, to prevent illegal setups.
- The form, color and material of courtyard wall shall be harmonic with the house and surrounding environment, and it is encouraged to beautify the walls according to local conditions to demonstrate the cultural feature of the community.
- Unified anti-falling canopy can be added when conditions permit.
- Canopy can be added at the courtyard entrance, to enhance the labeling.
- The entrance space can be beautified by adding green landscape and decoration details.

图2-1-22 院墙高度兼顾视线和景观要求 / Both sight and landscape requirements shall be considered for wall height | 图片来源：编写组自摄

图2-1-23 院墙装饰美化 / Courtyard walls are decorated for good appearance | 图片来源：编写组自摄

图2-1-24 底层院落入口加强细节设计 / Enhanced detail design for courtyard entrance | 图片来源：编写组自摄

1.3 改善建筑外观形象
Improve building appearance

图2-1-25 南京金尧山庄小区单元入口改造 / Unit entrance renovation in Jinrao Villa Community of Nanjing | 图片来源：编写组自摄

▶ 一体化改造单元入口空间
Integrated renovation of unit entrance space

· 将入口雨篷、踏步、坡道、门禁系统、单元编号以及休息座椅等要素整体统筹设计，营造具有细节感、人性化的入口空间。

· Integrated design shall be made for entrance canopy, steps, ramp, access control, unit numbering and rest chairs, to create humanized entrance space with sense details.

图2-1-26 单元入口空间结合加装电梯整体改造 / Overall renovation of unit entrance space plus adding elevator | 图片来源：编写组自摄

图2-1-27 单元门及门禁系统 / Unit door and access control system | 图片来源：编写组自摄

图2-1-28 单元门禁系统 / Unit access control system | 图片来源：编写组自摄

图2-1-29 楼栋标识 / Building marking | 图片来源：编写组自摄

图2-1-30 单元门整体更换 / Replace the whole unit door | 图片来源：编写组自摄

图2-1-31 宅间居民休息设施 / Rest facilities for residents | 图片来源：编写组自摄

南京天顺苑楼栋单元入口改造：
单元入口结合加装电梯整体设计，增加了入口雨篷，更换了性能更好的单元门和门禁系统，也增加了无障碍坡道等适老便民设施。

Building unit entrance renovation in Tianshun Garden of Nanjing:
Overall design of unit entrance combined with adding elevator, entrance canopy was added, replaced with unit door and access control system with better performance, also added with barrier-free ramp and other facilities convenient to the elderly.

● 单元门整体更换　● 单元编号　● 单元门禁系统
● 无障碍坡道　● 加装电梯　● 入口雨篷

不同程度提升所在地区风貌特色

Upgrade the stylistic feature of the area to different extent

- 与周边环境协调，通过色彩、材质等设计呼应周边环境。
- Harmonic with the surrounding environment, to echo with the surrounding environment with design of colors and materials.

昆山琼花新村建筑外观整治：
地处昆山老城区，建筑改造通过黑白灰色彩、墙面装饰线条、木色格栅构架等设计手法，实现与周边传统环境相协调。

Building appearance improvement in Qionghua New Village of Kunshan:
It is located in the old urban area of Kunshan, design means of using the black, white and gray colors, decorative lines on wall and wooden color grid are used in the renovation, to be harmonic with the surrounding traditional environment.

图2-1-32 昆山琼花新村呼应老城江南传统风格 / Qionghua New Village in Kunshan echoing the traditional Jiangnan style in the old town | 图片来源：编写组自摄

- 挖掘所在地区的建筑文化，进一步提升地区整体风貌特色。
- Tap the architectural culture of the area, to further improve the overall feature of the area.

南京宁夏路13号建筑外观整治：
地处颐和路民国建筑风貌区，通过仿小灰砖的外墙涂料、具有民国特色的装饰线脚等细部设计手法，融入并提升了地区的整体风貌特色。

Building appearance improvement at 13 Ningxia Road in Nanjing:
It is located in the Republic of China architectural style area of Yihe road, the building is merged into the environment by detailed design of exterior wall painting imitating small gray bricks and featured surbases, and the stylistic feature of the whole area has been improved.

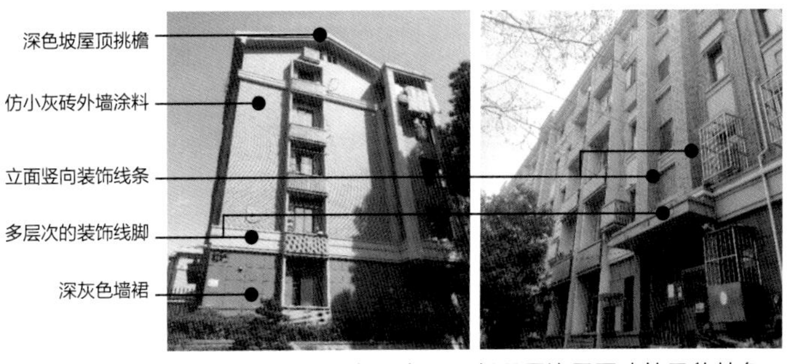

图2-1-33 南京宁夏路13号彰显周边民国建筑风貌特色 / 13 Ningxia Road in Nanjing, demonstrating the Republic of China architectural feature in surrounding area | 图片来源：编写组自摄

广州恩宁路永庆坊综合改造：
地处广州最具特色的西关旧址区域，项目改造采取原地升级的做法，融“新”于“旧”，通过运用瓦屋面、青砖墙、红砖墙、趟栊门等地方传统建筑元素，延续岭南民居特色，保持并进一步强化了所在地区的传统风貌和文化特色。

Comprehensive renovation of Yong Qing Fang at Enning Road in Guangzhou:
It is located in the most featured old site of Xiguan in Guangzhou, the renovation is upgrading from the original by merging the “new” into the “old”, the local traditional architectural elements such as tiled roof, black and red brick walls and Tanglong door, to extend the features of civil dwellings in Lingnan and maintain and further enhance the traditional style and cultural features of the area.

图2-1-34 广州恩宁路永庆坊改造延续西关风情特色 / Renovation of Yong Qing Fang at Enning Road in Guangzhou extends the style and feature of Xiguan | 图片来源：编写组自摄

1.4 提高节能性能
Raise energy saving performance

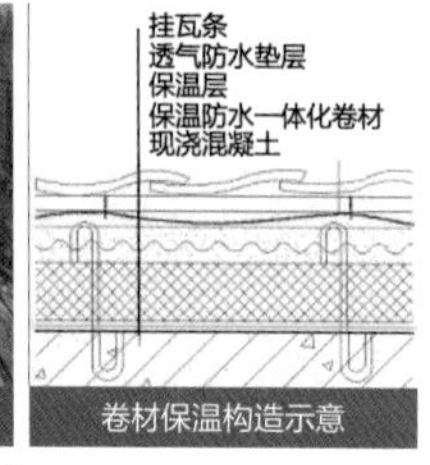

图2-1-35　提高建筑外墙保温的措施 / Measures for better insulation of building exterior wall | 图片来源：编写组自摄、自绘

图2-1-36　顶层露台采用中空节能玻璃 / Hollow energy saving glass for roof balcony | 图片来源：编写组自摄

图2-1-37　楼梯间采用节能玻璃 / Energy saving glass for staircase | 图片来源：编写组自摄

图2-1-38　封闭改造处理的楼栋单元门，增加保温性能 / Building unit door renovated by enclosing for better insulation | 图片来源：编写组自摄

建筑围护结构保温改造
Insulation renovation for building cladding

- **屋顶及外墙保温：**可通过增加保温板、保温涂料及铺设保温卷材等方式提高保温效果。
- **Roof and exterior wall insulation:** heat preservation effect can be improved by adding insulation board, insulation coating and laying insulation sheets.

- **外窗保温：**窗框质量良好时，优先更换玻璃部分，采用中空的节能玻璃；对于窗外框质量较差的，采用整体更换的方式。
- **External window insulation:** when the window frame is of good quality, the glass shall be replaced first with hollow energy-saving glass; overall replacement shall be made for window external frames in poor quality.

- **楼栋单元门保温：**应选用集保温隔热、防火、防盗等功能于一体的单元门，如原有通透式楼宇门较好的情况下，可对通透部分用安全玻璃进行封闭改造。
- **Building unit door insulation:** unit door with functions of thermal insulation, fire and theft prevention shall be used, when the original hollow building door is in good condition, the hollow part can be enclosed with safety glass as renovation.

- **节能改造注意事项：**既有建筑实施节能改造前，应先进行节能诊断与评估，根据节能诊断与评估结果，再制定节能改造方案；保温改造宜与屋面防水、外立面改造同时进行，减少施工投入；节能改造中材料的性能、构造措施、施工要求应符合相关技术标准要求。
- **Precautions in energy saving renovation:** before the renovation of existing buildings, energy saving diagnosis and evaluation shall be made first, to work out the renovation scheme according to the diagnosis and evaluation results; the insulation renovation should be made concurrently with the renovation of roof water-proof and exterior facade, to reduce construction investment; the material performance, structural measures and construction requirements in energy saving renovation shall comply with the requirements of relevant technical standards.

推广运用低能耗设备
Popularize the use of low energy consumption equipment

- 建筑入口门厅、走廊、楼道采用节能灯具（LED），开关控制方式采用声控或光控。
- 公共区域节水器具改造，可在洗手间洗手台加装红外感应式装置，在便池采用脚踏延时出水阀，有效防止人为疏忽导致浪费的现象发生。

- Building entrance halls and corridors adopt energy-saving lamps (LED), to be turned on or off with sound control or light control.
- For water saving device renovation in public areas, infra-red inducing devices can be provided at washing basins and pedal water outlet valve with time delay provided at urinals in toilets, to effectively prevent waste due to negligence by people.

图2-1-39 屋顶太阳能提供公共车库LED照明 / Roof solar energy device supply LED lighting in public garage | 图片来源：编写组自摄

太阳能光电系统+LED照明：
太阳能非逆变LED照明（PV-LED）技术是指将太阳能光伏发电融入建筑一体化中，采用高效智能控制技术，将组件-控制-并网-储能-LED灯具，构建成一个发电用电的直流系统，以光伏电力解决建筑内公共区域的照明问题，以达到节能目的。

Solar energy photoelectrical system + LED lighting:
With solar energy non-inverted LED lighting (PV-LED) technology, the solar energy PV power generation is integrated into building integration, and high efficiency intelligent control technology is adopted, to integrate the assembly-control-grid connection-energy storage-LED fixtures into a power generation and consumption DC system, to provide lighting in public area with PV power for energy saving purpose.

图2-1-40 屋顶安放太阳能光伏发电设备 / Solar energy PV power generation equipment on roof | 图片来源：编写组自摄

1.4 提高节能性能
Raise energy saving performance

图2-1-41　地源热泵技术示意图（夜间）/ Schematic diagram for ground-source heat pump technology (night) | 图片来源：编写组自绘

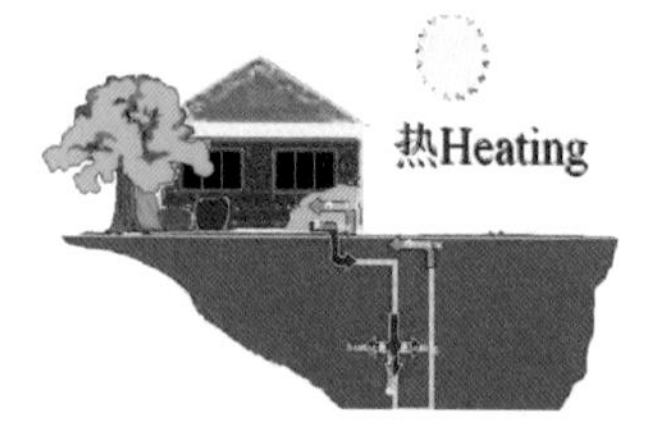

图2-1-42　地源热泵技术示意图（日间）/ Schematic diagram for ground-source heat pump technology (daytime) | 图片来源：编写组自绘

水（地）源热泵系统：
利用地球表面浅层水源（地下水、河流、湖泊）或者土壤中吸收的太阳能和地热能而形成的低位热能资源，并采用热泵原理，通过少量的高位电能输入，实现低位热能向高位热能转移。

Water (ground) source heat pump system:
The low-level heat energy resource formed by the solar energy and geothermal energy absorbed by the shallow water (ground water, rivers and lakes) or soil in the earth surface is used with heat pumps, to realize transfer of low-level heat energy to high-level heat energy with the input of small amount of high-level electric energy.

鼓励可再生能源利用
Encourage the use of renewable energy

- 结合屋顶空间条件，采用太阳能热水系统，减少燃气或电的消耗。
- 公共区域照明采用太阳能光电系统。
- 地源热泵系统有较好的节能效果，也能提升居住舒适度，但对既有建筑的改造要求较高，应在加装前调研其可行性。

- Solar energy hot water system can be used on the roof, to reduce consumption of gas or power.
- Solar energy PV system can be used for lighting in public areas.
- Ground-source heat pump system has good energy saving effect and can also make dwelling more comfortable, but it has high requirements on the renovation of existing buildings, therefore its feasibility shall be investigated before application.

图2-1-43　无负压节能供水设备 / Energy saving water supply equipment without negative pressure | 图片来源：编写组自摄

图2-1-44　节水型储水设备 / Water storage equipment with water saving feature | 图片来源：编写组自摄

图2-1-45　节水型自动灌溉设备 / Water saving automatic irrigationequipment | 图片来源：编写组自摄

图2-1-46　户外太阳能路灯 / Outdoor solar energy road lamps | 图片来源：编写组自摄

图2-1-47　节能草坪灯 / Energy saving lawn lamps | 图片来源：编写组自摄

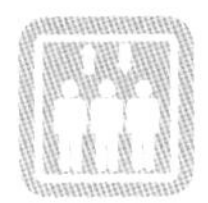

1.5 鼓励加装电梯
Encourage adding elevators

加装电梯适用建筑范围
Scope of buildings suitable for adding elevators

· 申请加装电梯的建筑应为建成投入使用、具有合法权属证明、未列入房屋征收改造计划且未设电梯的四层及以上非单一产权住宅。

· A building requesting for adding elevator shall be a residence building of four storeys and higher not owned by a single owner, already completed in use, with legal ownership certificates, not included in house expropriation and renovation plan and without elevator.

遵循地方加装电梯的相关规定
Relevant local stipulations on adding elevators shall be followed

· 前提条件：一般应满足楼栋单元户数的一定比例，并需对费用分担等相关事项达成协议，具体应遵循地方相关规定要求。

· Prerequisite: generally it shall be agreed by a certain proportion of households in a unit, agreement will be made for matters such as expense sharing, and relevant stipulations of local area shall be followed.

图2-1-48 南京天顺苑北侧错层入户 / Tianshun Garden in Nanjing Access between floors on north side | 图片来源：编写组自摄

图2-1-49 扬州油田社区南侧平层入户 / Yangzhou Oilfield Community Floor level access on south side | 图片来源：编写组自摄

加装电梯的一般流程
General process for adding elevator

· 按照各地方政府发布的既有建筑加装电梯办法或服务指南要求。一般包括：

第一步：业主意见征询

第二步：方案设计

第三步：现场公示

第四步：规划许可办理

第五步：施工许可办理（不具备办理规划许可、施工许可的可以采用部门联合会审形式办理）

第六步：视情况办理消防设计备案、人防改造手续

第七步：电梯安装质监施工公告

第八步：施工安装

第九步：电梯验收

第十步：交付使用

· Methods or service guide requirements for adding elevator in existing building issued by local governments shall be followed. Generally they include:

Step 1: solicit opinions from owners

Step 2: scheme design

Step 3: spot publicity

Step 4: obtain planning permit

Step 5: obtain construction permit(For those not qualified for planning permit and construction permit, joint review by departments can be arranged)

Step 6: go through procedures for fire protection design and air defence renovation as the case may be

Step 7: quality supervision and construction announcement for elevator installation

Step 8: construction and installation

Step 9: acceptance

Step 10: delivery for service

1.5 鼓励加装电梯
Encourage adding elevators

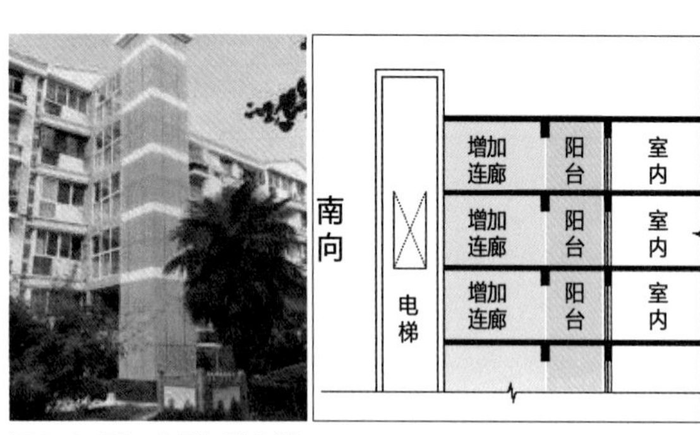

优点：从南侧阳台处平层进入，便捷性较高
缺点：电梯南向布置，会产生日照遮挡

Advantage: more convenient with access on the floor level at terrace on the south side
Disadvantage: elevator on the south side will affect the sunshine

图2-1-50　南侧加装电梯 / Elevator installed on the south side | 图片来源：编写组自摄、自绘

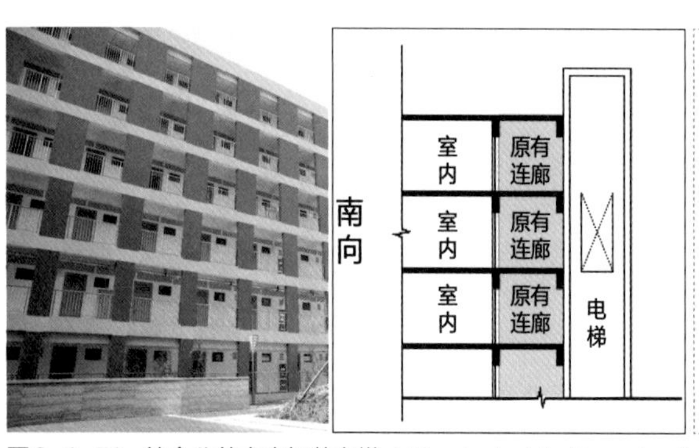

优点：结合公共走廊设置，加装改动和投入少
缺点：只能针对特定房型使用

Advantage: less change and investment when it is located in conjunction with the public corridor
Disadvantage: only applicable to specific forms of buildings

图2-1-51　结合公共走廊加装电梯 / Elevator installed in conjunction with the public corridor | 图片来源：编写组自摄、自绘

优点：结合楼梯间布置，对建筑及周边影响较小
缺点：连接楼梯半层平台，需上下半层，便捷性不足

Advantage: arranged in combination with the staircase, with less influence on the building and surrounding
Disadvantage: it is connected with the half-storey platform of the stairs, not quite convenient as walking up or down by half storey is required

图2-1-52　北侧加装电梯 / Elevator installed on the north side | 图片来源：编写组自摄、自绘

▶ 因地制宜选择合适的加装方案
Select suitable schemes according to actual conditions

- 电梯位置：一般建议布置在建筑北立面，如有必要可考虑山墙和建筑南立面，应尽量减少对现有住宅采光、通风等影响。
- 电梯入户方式：分平层入户与错层入户两种方式，结合楼栋现有条件、居民意愿、资金投入等合理选择。
- 加装电梯以实用为原则，选择可操作性强，便于实施的电梯形式，控制体量，尽量减少对本楼及周边建筑的消极影响。
- 老旧小区改造应考虑预留加装电梯空间，地下管线和地面设施的布设应为后续加装电梯做好基础。

- Elevator location: normally it is suggested to locate it on the north side of the building, when necessary, gable side or south side can be considered, and the influence on the lighting and ventilation of existing building shall be minimized.
- House access forms: either on the floor level or between floors, to be rationally selected according to exiting conditions of the building, desires of residents and investment.
- Elevators shall be added in the principle of practicability, elevators easy for operation and implementation shall be selected, and size be controlled, to minimize the negative effect on the own building and surrounding buildings.
- In the renovation of old communities, space for adding elevator shall be reserved, and the layout of underground pipelines and ground facilities shall lay a good foundation for subsequent adding of elevator.

电梯外观应与周边环境相协调
The elevator appearance shall be harmonic with the surrounding environment

- 同小区、同一幢住宅加装电梯应在结构形式、材质、风格等方面尽量一致，且与周边既有建筑相协调。
- Elevators in the same living quarter and building shall be identical in structural form, materials and style as far as possible, and also be harmonic with existing surrounding buildings.

南京五福家园小区5幢1单元加装电梯：
· 小区背景：南京市五福家园是拆迁安置房小区，老龄化十分严重，60岁以上的老人超过25%，上下楼成为老年居民的大问题。
· 加装过程：①通过自发协商取得了全单元居民书面同意。②委托南京市第二建筑设计院进行勘察设计。③形成初步方案报规划部门初审并公示。④公示后，携相关材料报规划部门，办理建设工程规划许可证。⑤单元居民与施工单位（江苏共创建设工程有限公司）、监理单位（南京旭光建设监理有限公司）签订合同后，向建设部门提交相关材料，办理施工许可证。⑥单元居民携相关设计材料，向消防机构办理消防设计备案。⑦施工单位书面告知区特种设备安全监督管理部门。⑧施工单位进场施工。⑨单元居民组织竣工验收，并向建设、人防、消防部门申请竣工验收备案。同时，向区特种设备安全监督管理部门办理使用登记，取得使用登记证书。⑩电梯加装完毕，正式使用。

Elevator added to Unit 1, Building 5 in Wufu Garden of Nanjing:
· Background: this living quarter is for settlement of relocated residents, with old people over 60 years over 25%, so it is a big problem for old people to go up and down stairs.
· Procedures: ① Written agreement of residents of the whole unit through spontaneous consultation. ② Entrusting Nanjing No. 2 Architecture Design Institute for survey and design. ③ Prepare the preliminary scheme to be submitted to the planning authority for preliminary review and publicity. ④ After publicity, obtaining construction project planning permit from the planning authority with relevant documents. ⑤ After the residents signed contract with construction contractor (Jiangsu Gongchuang Construction Engineering Co., Ltd.) and supervisor (Nanjing Xuguang Construction Supervision Co., Ltd.), relevant documents were submitted to the construction authority to obtain the construction permit. ⑥ Residents submit the relevant design documents to the fire protection authority for filing on fire protection design. ⑦ The construction contractor informed in written form the special equipment safety supervision and administration department of the district. ⑧ The construction contractor entered the site for installation. ⑨ The residents organized completion acceptance, and applied for completion application and filing with the construction, air defence and fire protection authorities. In the meantime, registration for use was done with the special equipment safety supervision and administration department of the district, to obtain the registration certificate for use. ⑩ The elevator installation is completed for service.

图2-1-53 玻璃材质弱化存在 / Glass may become weakened | 图片来源：编写组自摄

图2-1-54 立面材质与原有建筑协调 / Facade materials harmonic with existing building | 图片来源：编写组自摄

Construction Guidance for
Renovation of Old Communities in
Jiangsu

江苏老旧小区改造建设导引

2

消除安全隐患
Eliminate Security Risks

- 2.1 保障消防系统的安全运行
  Ensure safe operation of fire protection systems
- 2.2 完善安防设施保障居民出入安全
  Complete security facilities to safeguard the residents
- 2.3 满足应急防灾管理需要
  Meet the needs in emergency and anti-disaster management
- 2.4 建设公共设备智能监控系统
  Intelligent monitoring systems for public equipment in buildings
- 2.5 建设家居安防应急联动系统
  Provide security protection and emergency linked operation systems for residences
- 2.6 完善小区及周边疏散避难设施
  Improve evacuation and shielding facilities in communities and surrounding areas

2.1 保障消防系统的安全运行
Ensure safe operation of fire protection systems

依据标准规范完善消防设施
Complete firefighting facilities according to standards and specifications

- **室外消火栓：** 结合消防通道布设，应保证每幢建筑都在其保护半径以内；*DN*100或*DN*150的主出水口应朝向消防车道，便于消防车取水；杜绝将室外消火栓设在绿岛中央或无消防车道处的情况。
- **室内消火栓：** 超过7层的单元式住宅，超过6层的塔式住宅、通廊式住宅、底层设有商业网点的单元式住宅及所有的高层住宅应设置室内消火栓系统。
- **水泵接合器：** 水泵接合器应设在消防车便于操作的地点，距水泵接合器15~40m范围内应设置室外消火栓。
- **灭火器：** 住宅每层的公共部位建筑面积超过100m^{2}时，应配置1具1A的手提式灭火器；每增加100m^{2}时，增配1具1A的手提式灭火器；每个非机动车充电桩区域应配置1具1A的手提式灭火器。
- **微型消防站：** 合理划分最小灭火单元，配建微型消防站，配备消防器材及人员。
- **维护管理：** 定期检修和维护消防设施，且进行完整记录。

- **Outdoor hydrants:** in combination with the layout of fire passages, each building should be guaranteed within the protection radius; the main *DN*100 or *DN*150 water outlets should face the fire engine lane to facilitate fire engines to get water; avoid locating the outdoor hydrants in the middle of green island or in areas without fire engine lanes.
- **Indoor hydrants:** indoor hydrant system shall be installed in unit residential buildings with more than 7 storeys, tower residential buildings with more than 6 storeys, corridor residential buildings, unit residential buildings with commercial outlets on the ground floor and all high-rise residential buildings.
- **Pump adapters:** pump adapters should be located in places convenient for the operation of fire engines. Outdoor hydrants should be installed within 15~40m from pump adapters.
- **Fire extinguishers:** one 1A portable fire extinguisher shall be provided when the building area of public parts on each floor of the residence exceeds 100m^2; one 1A portable fire extinguisher shall be added for every more 100m^2; each non-motor vehicle charging stake area shall be provided with one 1A portable fire extinguisher.
- **Miniature fire station:** rationally divide the smallest fire extinguishing units, build miniature fire stations and provide firefighting equipment and personnel.
- **Maintenance and management:** firefighting facilities shall be regularly maintained and repaired, and complete records shall be made.

图2-2-1 消火栓远离消防通道 / Fire hydrants far from fire access | 图片来源：编写组自摄

图2-2-2 消火栓靠近消防通道 / Fire hydrants close to fire access | 图片来源：编写组自摄

加强消防通道安全管理
Strengthen safety management for fire protection passages

- 结合小区道路系统，完善消防车道规划设计，保证消防车道满足4m净宽，车道转弯半径不小于6m。
- 老城街巷无法满足车道4m的情况下，通过采取消防摩托车、微型消防车通道方式，满足消防安全需求。
- 环形消防车道至少应有两处与其他车道连通。尽端式消防车道应设置回车道或回车场，回车场的面积不应小于12m×12m；对于高层建筑，不宜小于15m×15m；供重型消防车使用时，不宜小于18m×18m。
- 消防车道的路面、救援操作场地及其下面的管道和暗沟等，应能承受重型消防车的压力。
- 加强消防通道管理，设置标识，保持消防通道畅通，严禁私自占用、堵塞、违章破路。

- Improve the planning and design of fire lanes in combination with the community road system, to ensure that the fire lanes are 4m in net width, with turning radius no less than 6m.
- In old towns where the 4m width cannot be met for streets and lanes, firefighting motorcycles and miniature fire engine passage shall be used to meet the fire protection safety requirements.
- A ring fire engine lane shall be connected with other lanes at no less than two places. A terminal type fire lane shall be provided with a return lane or return ground, and the area of return ground shall not be less than 12m×12m; for high-rise buildings, it should not be less than 15m×15m; when used for heavy fire engines, it should not be less than 18m×18m.
- The road surface of fire lanes, rescue operation ground and the pipes and drains beneath shall be able to withstand the pressure of heavy fire engines.
- Strengthen the management of fire accesses, set up signs, keep the fire accesses unblocked, prohibit unauthorized occupation, blocking, and road breaking.

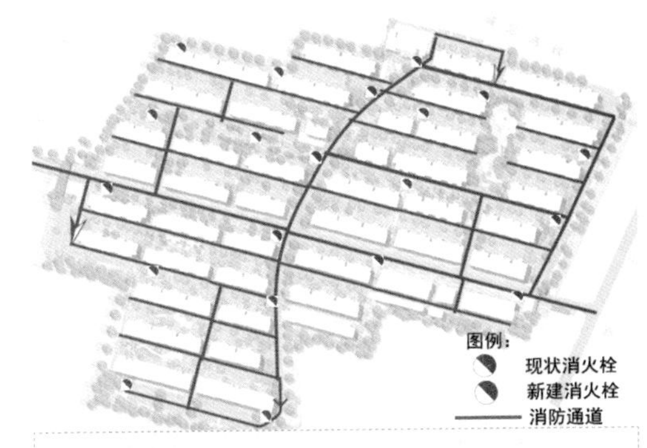

- 调研现状消火栓，评估现状消防能力。
- 补充设置消火栓，保护面积应覆盖整个小区。
- Investigate status quo hydrants and evaluate the status quo firefighting capability.
- Replenish fire hydrants, and the protection area shall cover the whole community.

图2-2-3 依据规范补充室外消火栓 / Replenish outdoor hydrants according to codes | 图片来源：枫景苑A区改造成果

图2-2-4 消防通道标识 / Fire access signs | 图片来源：编写组自摄

2.1 保障消防系统的安全运行
Ensure safe operation of fire protection systems

设置清晰的安全疏散指示标识
Set up clear safety evacuation indications and signs

· 安全疏散指示包括避难场所位置、逃生通道路径、安全出口等，指示标识及设置场所应满足相关标准规范。

· 各类安全指示内容应通俗易懂，作图规范，图文清晰。

· Safety evacuation indications include shelter venues, escape passage and path, safety exit, etc., indicating signs and venues should meet the relevant standards and codes.

· Contents of all kinds of safety indication should be easy to understand, pictures should be formalized, and both pictures and wording shall be clear.

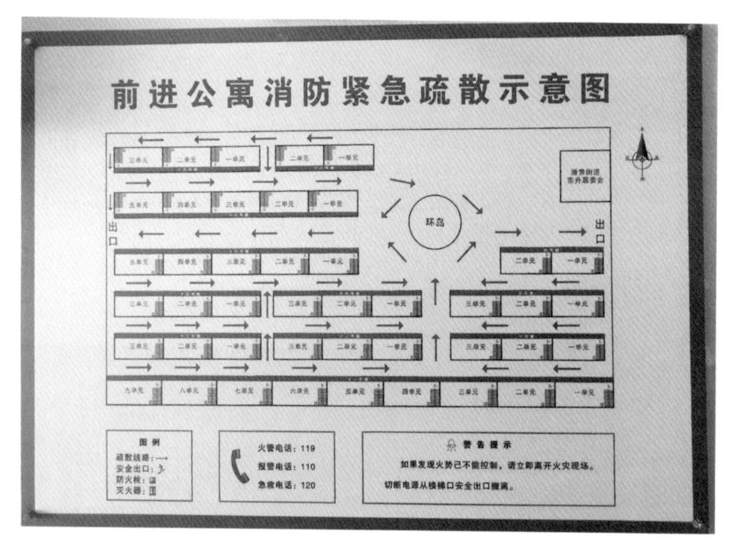

图2-2-5 公共建筑安全疏散指示图 / Safety evacuation maps for public buildings |
图片来源：编写组自摄

建立应急响应机制
Establish emergency response mechanism

- **制定预防突发机制：**物业应制定小区预防突发机制，内容包括消防设施检查维护、消防通道管理、消防安全培训、组织疏散练习，明确各方责任。
- **加强消防安全培训：**街道社区等相关部门会同消防管理机构每年至少组织一次对小区消防安全管理人员应对突发事件的预防、应急处置及协调等相关知识的培训，增强消防安全管理人员应对突发事件的处置能力。
- **开展消防知识宣传：**消防部门利用多种渠道和方式每年开展一次小区居民消防知识的宣传，做好安全教育进社区工作，增强居民的防范意识。
- **定期组织疏散演习：**物业应每年开展一至两次应急疏散演习，熟悉应急工作的指挥、协调和处置程序，不断完善应急预案。

图2-2-6 消防宣传 / Fire protection publicity | 图片来源：编写组自摄

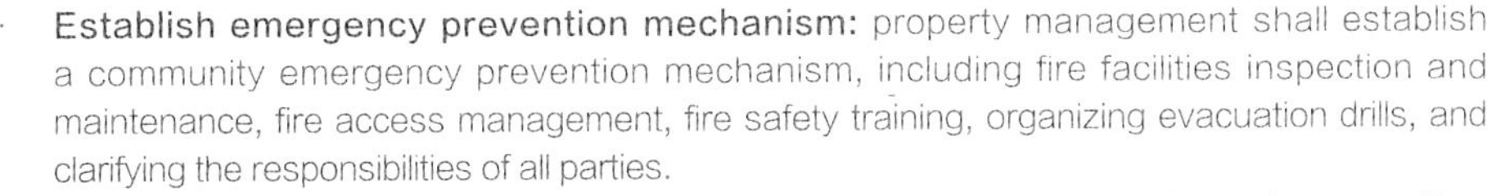

- **Establish emergency prevention mechanism:** property management shall establish a community emergency prevention mechanism, including fire facilities inspection and maintenance, fire access management, fire safety training, organizing evacuation drills, and clarifying the responsibilities of all parties.
- **Strengthen fire safety training:** relevant departments such as subdistrict and communities, together with fire management institutions, shall organize training of fire safety management personnel in the community at least once a year on relevant knowledge in the prevention, emergency response and coordination of emergency incidents, so as to enhance the ability of fire safety management personnel in dealing with emergencies.
- **Fire protection knowledge publicity:** the fire protection authority shall use various channels and methods to publicize fire protection knowledge on residents in the community once a year, to ensure safety education into the community and enhance the awareness of prevention of residents.
- **Regularly organize evacuation drills:** the property management shall organize one or two emergency evacuation drills every year, to get familiar with the command, coordination and handling procedures of emergency, and constantly improve emergency plans.

图2-2-7 消防培训 / Fire protection training | 图片来源：编写组自摄

图2-2-8 消防演习 / Firefighting drills | 图片来源：编写组自摄

2.2 完善安防设施保障居民出入安全
Complete security facilities to safeguard the residents

图2-2-9 小区出入口增设智能门禁系统 / Add intelligent access control system at community entrances | 图片来源：编写组自摄

图2-2-10 结合物业用房增设视频监控中心 / Add video monitoring center in property management room | 图片来源：编写组自摄

安防系统覆盖范围
Coverage of security system

· 包括小区周界、出入口、道路、公共区域以及建筑楼道、通道等公共部位。

· Covering the community perimeter, accesses, roads, public areas and building stairways, passageway and other public parts.

图2-2-11 小区出入口安防设施 / Security facilities at community entrance | 图片来源：编写组自摄、自绘

增设出入口智能门禁系统
Add intelligent access control system at entrances

· 小区出入口应能实施封闭管理，配设门卫值班室；鼓励采用车辆管理、人脸识别等智能门禁系统，闸口设置与城市道路应有一定缓冲距离，不影响城市道路交通。

· 楼栋单元入口应增设防盗门和门禁系统，要求安装牢固，并与建筑风格相协调。

· It shall be possible to implement enclosed management for community entrances, with guard duty room provided; intelligent access control systems, such as vehicle management and face recognition, are encouraged. Certain buffer distance should be set between the gate and urban road, without affecting the urban road traffic.

· The entrance of the building unit should be provided with security door and access control system, which should be firmly installed and coordinated with the architectural style.

完善公共区域安防监控系统
Complete security monitoring system in public areas

- 小区各主要出入口、周界围墙、道路、停车场、公共活动场所等区域安装监控摄像头，不留监控盲点。
- 结合物业管理中心或门卫设置视频监控中心和相应的辅助设施。

- Monitoring cameras shall be installed in all main entrances of the community, perimeter fence, roads, parking lots, public activity venues and other areas, leaving no blind spots for monitoring.
- Video monitoring center and corresponding auxiliary facilities shall be set up in conjunction with property management center or gate guard.

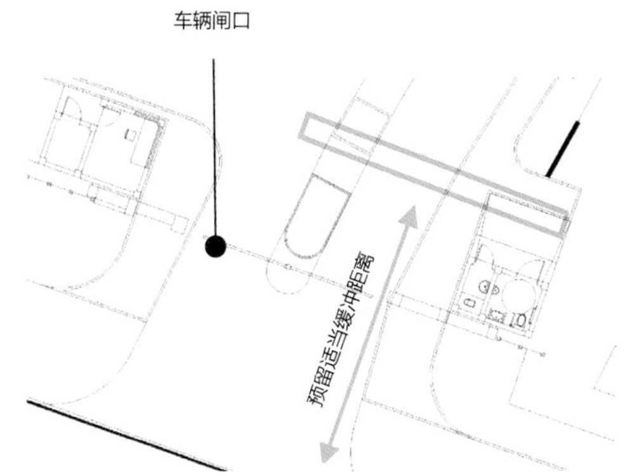

图2-2-12 小区出入口安防设施布局图 / Layout of security facilities at community entrance | 图片来源：编写组自绘

建立多层次的安全保障巡逻系统
Establish multi-level security guard patrol system

- 物业管理公司应配备专业安保队伍，制定日常巡护制度和安保措施，定时巡逻及时解决小区安防问题。
- 发动居民组织，作为安保补充力量，筑牢安全防线。

- A property management company shall be staffed with a professional security team, formulate the daily patrol system and security measures, and regularly patrol and promptly solve security problems in the community.
- Organize the residents, as a supplement force to security, to consolidate the security line.

图2-2-13 单元入口门禁设施布局图 / Layout of access control facilities at unit entrance | 图片来源：编写组自摄、自绘

2.3 满足应急防灾管理需要
Meet the needs in emergency and anti-disaster management

合理组织小区入口空间方便应急管理
Rationally organize community entrance space to facilitate emergency management

- 小区主要出入口应尽量实现进出分离、人车分流，便于应急状态下采取非常规措施时，不影响出入秩序。
- 出入口附属设施包括门卫、围墙、快递设施等应合理布局，减少流线交织，如快递设置，有条件的可考虑外投内取的方式，有利于外部人员管理。

- At community main entrances, the flow of people and vehicles should be separated as far as possible, to facilitate adopting unconventional measures in emergency without affecting the order of access.
- The auxiliary entrance facilities, including guards, fences and express delivery facilities, should be rationally arranged, to reduce the interweaving of flow lines. For example, external delivery and internal taking can be considered for express delivery when conditions permit, to facilitate management of external personnel.

预留一定弹性空间满足应急功能需要
Reserve some elastic space to meet the needs of emergency functions

- 老旧小区改造时，可结合出入口、交通方便的公共空间，预留一定的弹性空间，便于增设临时性功能或设施，满足应急状态下的需求。

- In the renovation of old communities, some elastic space can be reserved in combination with the public space with convenient access and traffic, so as to facilitate the addition of temporary functions or facilities to meet the needs under the emergency state.

图2-2-14 小区出入口便于应急管理 / Facilitate emergency management at community entrance | 图片来源：枫景苑A区改造成果

开展多样化的防灾教育活动
Carry out diversified educational activities on disaster prevention

- 结合国家防灾减灾日、国际减灾日等重大纪念日，开展防灾减灾宣传教育活动。
- 设置防灾减灾专栏，开展日常性的居民防灾减灾宣传教育。
- 定期邀请有关专家、专业人员或志愿者，对社区管理人员和居民进行防灾减灾培训。

- Publicity and education activities on disaster prevention and reduction shall be carried out in conjunction with major anniversaries such as the National Day for Disaster Prevention and Reduction and the International Day for Disaster Reduction.
- Set up special columns for disaster prevention and reduction, and carry out regular publicity and education on disaster prevention and reduction among residents.
- Regularly invite experts, professionals or volunteers to train community administrators and residents in disaster prevention and mitigation.

图2-2-15 小区防灾教育宣传展板 / Community disaster prevention education and publicity board | 图片来源：编写组自摄

提升社区居民减灾自救能力
Enhance the ability of community residents in disaster reduction and self-rescue

- 加强宣传，使居民清楚社区各类灾害风险及其分布，以及避难场所和疏散路径。
- 应使居民掌握防灾减灾自救互救基本方法与技能，了解救灾设施的使用方法。
- 积极创造条件，鼓励居民参与社区组织的各类防灾减灾活动，鼓励社区居民成立“灾害管理志愿者”组织，并提供场地进行防灾演练，确保灾害来临时，基层团体可以快速协助疏散居民。

- Strengthen publicity so that residents know clearly the risks and distribution of various disasters in their communities, as well as shelters and evacuation routes.
- Residents should master the basic methods and skills of disaster prevention and mitigation and self and mutual rescue, and know how to use disaster relief facilities.
- Actively create conditions to encourage residents to participate in various disaster prevention and mitigation activities organized by the community, encourage the residents to set up “disaster management volunteers” organizations and provide sites for disaster prevention drills, so as to ensure that grassroots groups can quickly help evacuate residents when disaster strikes.

2.4 建设公共设备智能监控系统
Intelligent monitoring systems for public equipment in buildings

▶ 智能消防系统
Intelligent fire protection system

- 接入城市智慧消防云平台，通过平台实现智能化的火灾隐患监测与防控。
- 将地理位置信息纳入智慧消防系统中，建设烟雾、用电、消防水源等监测系统，通过网络实现数据的实时上传、动态监测，如若出现紧急情况或警情信息，能够将准确的地址及相关情况数据传送至监控中心。
- 在老旧高层住宅、特殊人群住所等场所安装独立式烟感探测报警器、可燃气体探测器、温度传感器、智能喷淋等智能终端，构建智能隐患监测预警系统。
- 设置电气火灾报警装置，设施布置应符合现行国家标准和《火灾自动报警系统设计规范》GB 50116-2013的有关规定。

- It is connected to the urban intelligent fire protection cloud platform, for intelligent fire hazard monitoring and prevention and control via the platform.
- The geographic location information shall be incorporated into the intelligent fire protection system, and the monitoring system for smoke, electricity and firefighting water will be built. Real-time data uploading and dynamic monitoring will be realized via the network. In case of emergency or alarm information, accurate address and relevant data will be transmitted to the monitoring center.
- Independent smoke detector alarms, combustible gas detectors, temperature sensors, intelligent spray and other intelligent terminals can be installed in old high-rise residential buildings and special people's residences, to build an intelligent hidden danger monitoring and early warning system.
- Electrical fire alarm devices shall be set, and the arrangement of facilities shall comply with the current national standards *Code for design of automatic fire alarm system* GB 50116-2013 and relevant provisions.

▶ UPS 不间断电源系统
UPS power supply system

· 中心机房设置UPS后备电源，在市电发生故障时自动切换到UPS电源供电，保证设备的正常运行（后备时间：2小时）。

· UPS backup power supply shall be set up in the central machine room, so that power supply can be automatically switched to UPS when main power fails to ensure the normal operation of equipment (backup time: 2 hours).

图2-2-16 智能充电桩 / Intelligent charge stake | 图片来源：南京尧林仙居改造智能化系统方案成果

▶ 电动车智能充电系统
Electric vehicle intelligent charging system

· 电池充电完闭，设备自动切断电源，停止计时，杜绝火灾隐患。
· 系统自动识别电动车充电器，杜绝其他电器使用；如果用电功率超过电动车充电功率，则系统自动停止供电。
· 配备电子防盗锁，只要锁链断开，数据传输中断则通过报警装置报警。

· The equipment will automatically cut off power at the end of battery charge with timing stopped, to avoid hidden danger of fire.
· The system automatically recognizes the electric vehicle charger to prevent the use of other electrical appliances; if the power exceeds the charging power of the electric vehicle, the system will automatically stop the power supply.
· Electronic anti-theft lock is provided, when the lock chain is broken, data transmission is interrupted and alarm will be sent by the alarm device.

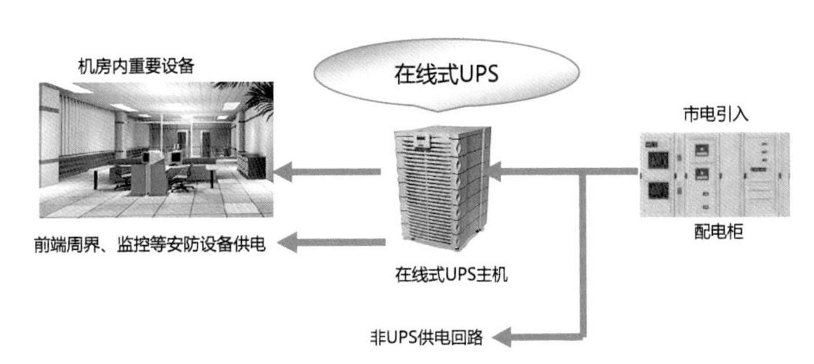

图2-2-17 UPS不间断电源系统 / UPS power supply system | 图片来源：南京尧林仙居改造智能化系统方案成果

2.5 建设家居安防应急联动系统
Provide security protection and emergency linked operation systems for residences

▶ 紧急按钮
Emergency pushbuttons

· 针对独居老人家庭，安装在家中可方便接触到的地方，当发生火灾或其他紧急情况时，通过一键触发按钮，经无线电通信设备进行拨号报警。

· Installed where easily accessible at home with single elderly, in case of fire or other emergency, the one-key trigger pushbutton can be used to dial and alarm via radio communication equipment.

▶ 警报推送
Alarm sending

· 用户设置多个不同的紧急联系方式，在触发不同报警类型时能联系到相应的人，报警装置通过网关将报警信息推送到用户手机。

· A user can set up a number of different emergency contacts. When different types of alarms are triggered, the corresponding person can be contacted. The alarm device can send the alarm message to the user's mobile phone via the gateway.

“SOS”和自动警报
“SOS” and automatic alarm

· 在门窗入口处安装智能传感器，探测非法入侵动作。当传感器被触发时，系统将引发报警装置发出鸣笛声，楼宇管理系统会将报警信号发送至物业中心，小区物业安保员可立即报警。

· Install smart sensors at the door and window entrances to detect illegal intrusion. When the sensor is triggered, the system will trigger the alarm device to sound its horn, and the building management system will send the alarm signal to the property center, so that the property security personnel of the community can immediately report the alarm.

安防警报管理
Security alarm management

· 在厨房安装烟雾传感器、煤气传感器、自动门窗和门磁传感器等。若户主离家，智能安防与联动系统将自动建立布防模式，各种探测器均时刻处于“备战”状态。

· Install smoke sensors, gas sensors, automatic door and window magnetic door sensors in the kitchen. When people left the house, the intelligent security and linkage system will automatically establish a defense mode, and all kinds of detectors are always in the “ready” state.

2.6 完善小区及周边疏散避难设施
Improve evacuation and shielding facilities in communities and surrounding areas

疏通避难救灾通道
Get through shelter and rescue passages

- 结合现有路网，疏通、连接街巷，完善消防、疏散、救灾通道，提高系统性。
- 整治疏散通道，移除障碍物，整理归并管线，保障通道畅通。

- Get through and connect streets and lanes in combination with the existing road network, and improve firefighting, evacuation and disaster relief channels, to make them more systematic.
- Rectify evacuation channels, remove obstacles, sort out and merge pipelines to ensure the unimpeded passage.

· 设置位置：小区楼梯间、大门出入口、社区公园等公共区域。
· Locations: public areas such as the community stairway, entrance gate, community park, etc.

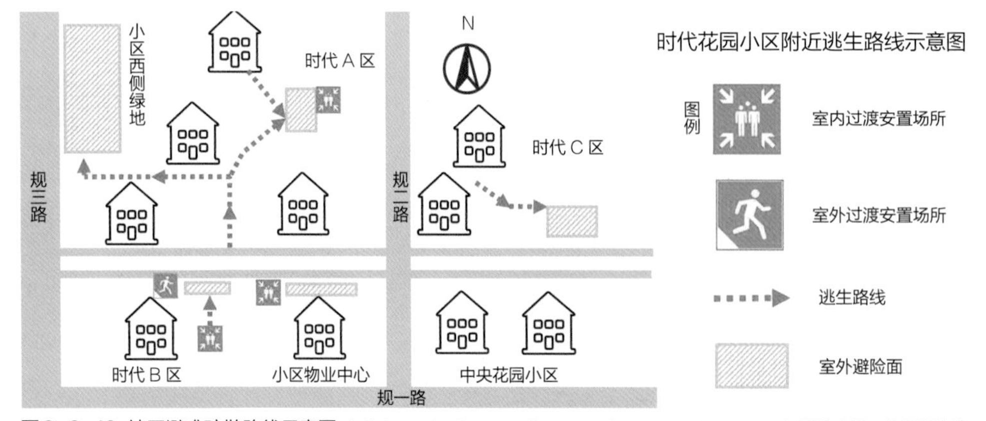

图2-2-18 社区避难疏散路线示意图 / Schematic diagram of community evacuation routes | 图片来源：编写组自绘

完善周边避难设施
Complete shelter facilities in surrounding areas

- 利用小区附近的公园、广场、绿地、空地以及抗震能力较高的公共设施，合理设置紧急避难场所与固定避难场所。
- 避难场所应设置引导性标示牌，牌上标出避难所的名称，绘制出责任区域的分布图和内部功能区分布图。

- Make use of the parks, squares, green space, open space and public facilities with high earthquake resistance in the vicinity of the community, to reasonably set up emergency shelter and fixed shelter venues.
- Guiding signs should be set up for shelter venues, with the name of shelter marked on the signs, and distribution maps of the responsibility areas and internal functional area should be plotted.

图2-2-19 社区灾害风险及周边避难设施指引图 / Guide map of community disaster risk and surrounding shelter facilities | 图片来源：编写组自摄

有条件的社区可建设防灾中心
Communities with the ready conditions can build disaster prevention centers

- 利用具有一定规模的社区公园（3000m^{2}以上，依据《城市社区应急避难场所建设标准》建标180-2017），建设防灾中心，配置相应防灾设施，满足应急避难功能需要。

- Community park with a certain size (over 3000m^2, according to the *Construction standard of urban community emergency shelter*, construction standard 180-2017) can be used, to build disaster prevention center and configure it with corresponding disaster prevention facilities to meet the needs of emergency shelter function.

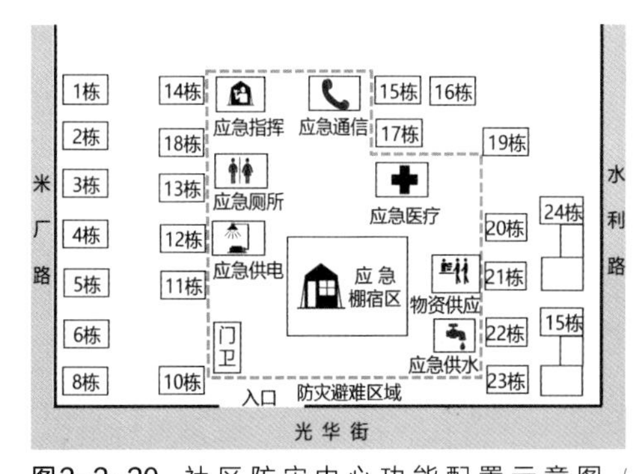

图2-2-20 社区防灾中心功能配置示意图 / Schematic diagram of community disaster prevention center function configuration | 图片来源：编写组自绘

Construction Guidance for
Renovation of Old Communities in
Jiangsu

江苏老旧小区改造建设导引

3

保障基础设施安全供应

Ensure the Safe Supply of Infrastructure

3.1 更新改造老化落后设施
Update and renovate aged and backward facilities

管网更新
Renewal of pipe networks

· 梳理各类管线的在用情况，对废弃管线予以清理。

· 检查各类管网漏损、管线老化、雨污混接、标准落后等影响使用功能或存在安全隐患的情况，对于破损、老化或不符合现行相关标准的管道、线路进行维修、更换。

· Sort out the conditions of all kinds of pipelines in use and remove the abandoned pipelines.

· Check all kinds of pipe networks for leakage, aging, rain and sewage water mixing, backward standards and other conditions that affect the use function or have hidden safety risks, repair or replace pipes and lines already damaged, aged or not complying with current standards.

图2-3-1 检查井防坠落设施 / Anti-falling facilities in inspection wells | 图片来源：编写组自摄

图2-3-2 更换破损井盖 / Maintain ventilation for inspection in the well | 图片来源：编写组自摄

设施维护
Maintenance of facilities

· 定期清洁、检修低位贮水池、高位水箱等供水设备，加强二次供水防污染措施。

· 排水检查井井盖，应符合强度等级标准，一般绿化带、人行道采用B125级，车行道采用C250级或D400级。

· 检查井井盖应有防盗、防沉降措施。

· 排水检查井内应有防坠落措施，如加装防坠网或防护内盖。

· Regularly clean and repair the low level water pools, high level water tanks and other water supply equipment, strengthen the pollution prevention measures for secondary water supply.

· Drainage inspection well covers shall conform to the strength class standards. Generally, they should be of Class B125 in green belts and sidewalks, and of class C250 or D400 for vehicle lanes.

· Provisions against theft and settlement shall be made for inspection well covers.

· Anti-falling provision shall be made in water drainage inspection wells, such as anti-falling net or protection inner cover.

完善安全警示标识
Complete safety warning signs

- 小区变电箱、燃气调压箱周边应设置警示标识。
- 敷设燃气管道地段需设置燃气桩或地标等明显的标识。
- Warning signs should be set around the substation and gas converter boxes in the community.
- Conspicuous signs such as gas stake or landmarks shall be set up in areas laid with gas pipelines.

图2-3-3 电箱警示标识 / Electrical box warning sign | 图片来源：编写组自摄

图2-3-4 燃气设施警示标识 / Gas facility warning sign | 图片来源：编写组自摄

图2-3-5 定期检查燃气设施 / Regular checking of gas facilities | 图片来源：编写组自摄

3.2 实施雨污分流改造
Renovation to separate rainwater and sewage water

图2-3-6 阳台立管改造前 / Balcony riser before renovation | 图片来源：编写组自摄、自绘

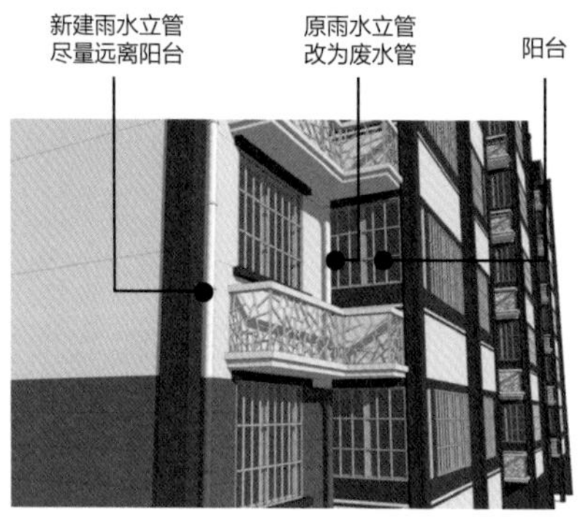

图2-3-7 阳台立管改造后 / Balcony riser after renovation | 图片来源：编写组自摄、自绘

图2-3-8 厨房立管改造前 / Kitchen riser before renovation | 图片来源：编写组自摄、自绘

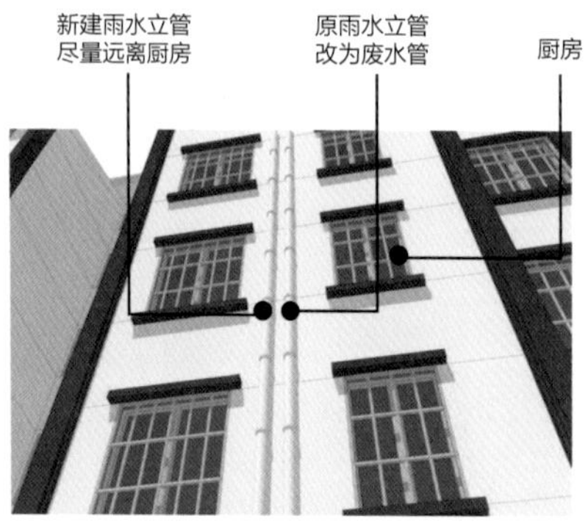

图2-3-9 厨房立管改造后 / Kitchen riser after renovation | 图片来源：编写组自摄、自绘

小区排水管线排查
Inspect water drainage pipelines in the community

- 通过检测雨水排口的旱流流量、氨氮浓度等方式，摸清混接情况。
- 必要时可通过CCTV、QV等可视化检测手段，判断管线完好程度。
- Make clear the mixing conditions by detecting the dry season flow and ammonia and nitrogen concentration at rainwater drain ports.
- If necessary, visual inspection means such as CCTV and QV can be used to determine the condition of pipeline.

制定管网改造方案
Work out pipe network renovation scheme

- 结合现状情况，校核污水量、雨水量，合理划分排水分区，制定雨污管网改造方案。
- 污水接户管靠近建筑敷设，雨水接户管、排水沟应远离建筑敷设，防止居民私接混接。
- Check the amount of sewage and rain water, reasonably divide the drainage areas, and formulate the rain water and sewage pipe network renovation plan according to status quo conditions.
- Sewage water pipes connecting with houses shall be laid close to building, and rainwater pipes and drain ditches shall be laid away from buildings, to prevent unauthorized mixed connection by residents.

改造混接立管
Renovate mixed risers

- 原立管可用作污、废水立管，接入污水井。
- 新敷设的雨水落水管需远离阳台、厨房、卫生间，防止居民再次混接。

- Existing risers can be used for sewage water and wastewater and be connected to sewage water wells.
- New rainwater downspouts shall be kept away from balconies, kitchens and toilets to prevent residents from mixing them again.

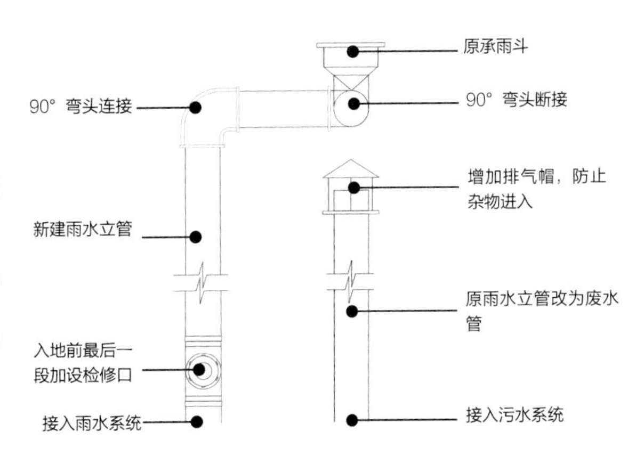

图2-3-10 雨水立管断接做法 / Rainwater riser disconnection and connection | 图片来源：编写组自绘

管网材料应满足快速化施工要求
Pipe network materials shall meet the requirements of quick construction

- 雨、污水管材根据埋深情况，采用塑料类管材或球墨铸铁管。
- 污水检查井采用塑料井、钢筋混凝土成品井。
- 雨水检查井采用砖砌井、塑料井、钢筋混凝土成品井。

- Plastic or nodular cast iron pipes shall be used for rainwater and sewage water pipes according to the burial depth.
- Sewage water inspection wells shall be made of plastics or finished reinforced concrete.
- Rain water inspection wells shall be made of bricks, plastics or finished reinforced concrete

图2-3-11 钢筋混凝土成品井 / Finished reinforced concrete well | 图片来源：编写组自摄

图2-3-12 塑料成品井 / Finished plastic well | 图片来源：编写组自摄

图2-3-13 砖砌井 / Brick well | 图片来源：编写组自摄

3.3 整治“三线”私拉私接现象
Rectify illegal connection of wires and pipes

“三线”整治原则
Principles for rectifying illegal connections

· 积极推进小区内供电、通信、有线电视等线路“三线入地”工程。
· 道路空间狭小、线路无法入地时，应对架空线路进行整理，保证架空线路整齐划一，保持横向水平，纵向垂直，统一靠墙设置，严禁随意搭接线路。

· Actively promote burying the power, communication and CCTV lines in ground in community.
· Overhead lines shall be re-arranged where the road space is narrow and it is not possible to bury the lines, to ensure that the overhead lines are neatly arranged, in good horizontal and vertical layout against walls, and it is strictly forbidden to connect lines arbitrarily.

“三线”入地
Bury the lines in ground

· 以下情况“三线”原则上应下地：
小区5m以上主要道路的架空线路；
横跨3m以上道路的架空线路；
具备下地条件的其他架空线路。
· 强电线路与弱电线路的最小水平和垂直净距均为0.5m，可通过以下方法缩小二者净距：
弱电线路附近平行敷设屏蔽线或弱电线路穿金属管道的方法进行电磁屏蔽保护；
使用电磁屏蔽电缆、电磁屏蔽塑料多孔管等新材料；
弱电线路采用不受电力线电磁场干扰的光缆。

· In principle, the lines shall be buried in ground in the following circumstances:
Overhead lines along main roads of 5m and over in the community;
Overhead lines crossing roads of 3m and over;
Other overhead lines with ready conditions for burying.
· The minimum horizontal and vertical net distance between strong and weak current lines is 0.5m, and the net distance between them can be reduced by the following methods:
Lay shielded lines parallelly by weak current lines or put weak current lines in metal conduits for electromagnetic shielding and protection;
Use electromagnetic shielding cables, electromagnetic shielding plastic porous tubes and other new materials;
Use optical cable not interfered by power line electromagnetic field for weak current lines.

“三线”规整
Line rectification

- 通过优化线路结构进行改造，按装饰性遮挡、入槽盒或线杆等方式进行有序规整，符合安全要求及横平竖直美观要求。
- 沿墙体敷设的强电线路可采用格栅或广告牌等方式遮蔽；弱电线路可采用槽盒或套管进行统一规整。
- 室外线路没有合适墙体敷设时，采用线杆规整方式，应进行集中束理、捆扎或套管处理。
- 强电线路应高于弱电线路，中压线路与弱电线路的垂直距离不小于2.5m，低压线路与弱电线路的垂直距离不小于1.5m。

- Renovate the lines by optimizing their structure, and put them in order by using decorative shading, putting into cable trays or on posts, to meet the safety requirements and appearance requirements.
- Strong current lines on walls can be shaded by grating or advertisement plates; weak current lines can be put into cable trays or sleeves for rectification.
- Where no suitable wall can be used to lay outdoor lines, they can be erected on posts, and should be well arranged, tied into bundles or put into sleeves together.
- Strong current line should be above the weak current line, the vertical distance between the MV line and weak current line should be no less than 2.5m, and the vertical distance between the LV line and weak current line should be no less than 1.5m.

架空线遮挡物选择
Selection of overhead line shading materials

- 格栅材质可选用铝合金、PVC以及木、木质复合等材质。
- 格栅样式可根据小区建筑风格选用现代、中式、西式等样式。
- 槽盒或套管可选用铝合金、PVC等材质。
- Aluminum alloy, PVC and wood or wood composite materials can be selected for gratings.
- Gratings can be in modern, Chinese and western styles according to the architectural style of the community.
- Aluminum alloy and PVC can be selected for cable trays or sleeves.

3.4 完善公共区域照明
Complete lighting in public areas

图2-3-14 庭 院 灯 / Courtyard light | 图片来源：编写组自摄

图2-3-15 草 坪 灯 / Lawn light | 图片来源：编写组自摄

公共区域照明要求
Lighting requirements for public areas

· 小区入口、道路、公共活动空间、停车场以及楼栋单元入口、楼道等公共区域应完善照明设施，满足居民夜间室外活动的需求并确保安全性。

· The community entrances, roads, public activity space, parking lot, entrance of building units, corridor and other public areas shall have complete lighting facilities to meet the needs of residents for outdoor activities at night and ensure safety.

照明设施选型
Selection of lighting facilities

· 小区主要考虑功能性照明，以庭院灯或普通路灯为主。结合公共活动空间，可适当设置草坪灯、投射灯等装饰性灯具，营造良好的夜景艺术效果。

· 小区入口大门可采用泛光灯及线性灯的照明方式，需注重灯具的隐藏。

· In a living quarter, functional lighting shall be mainly considered, with courtyard lights or ordinary street lights as the main. In public activity space, lawn lights, projection lights and other decorative lights can be set appropriately to create a good night artistic effect.

· At the entrance gate, floodlight and the linear lights can be used, and the light fixtures should be concealed.

防止照明设施产生光污染
Prevent light pollution from lighting facilities

· 合理安排灯具的间隔距离和位置，满足照明需要，同时不对住宅产生干扰光；光源宜采用高效节能灯具，如LED灯、太阳能灯具。

· 小区道路宜采用带遮光罩下照式庭院灯，照明光线严格控制在场地内，避免对周边住宅产生干扰光。

· Arrange the distance and position of light fixtures reasonably to meet the needs of lighting, without producing interference light to houses; the light source should be high efficiency energy-saving lamps such as LED lamps and solar lamps.

· Community roads should be illuminated with courtyard lamps with hood. The lighting light should be strictly controlled within the site to avoid interfering with the surrounding residences.

3.5 推进 5G 网络建设
Advance the construction of 5G network

加强科普宣传，为 5G 网络建设提供便利
Strengthen popularization of science to facilitate the construction of 5G network

- 加强科普宣传，消除公众对基站电磁辐射的片面认识，获得居民支持。
- 物业服务企业应为住宅小区5G网络建设提供便利，切实满足用户的5G需求，并为今后的5G+智慧物业搭建网络基础。

- Strengthen the popularization of science, eliminate the public's one-sided cognition of electromagnetic radiation of base stations, and obtain support from residents.
- Property service enterprises should provide convenience for 5G network construction in residential communities, effectively meet users' 5G demand, and build a network foundation for 5G+ smart property in the future.

图2-3-16 基站设施美化处理 / Beautify base station facilities | 图片来源：编写组自摄

5G 设施建设应符合小区整体美观要求
The construction of 5G facilities shall meet the overall aesthetic requirements of communities

- 5G基站优先结合现有基站设置，或结合小区公共建筑设置，设施建设应符合小区整体美观要求。
- 在不增大传播损耗的情况下，适当对天线等设备进行装饰，美化视觉效果。

- 5G base stations shall be set in combination with existing base stations or public buildings in the community, and the facility construction shall meet the overall aesthetic requirements of communities.
- The antenna and other equipment should be appropriately decorated without increasing the propagation loss, to get better visual effect.

图2-3-17 加强通信基础设施科普宣传 / Strengthen popularization of science for communication infrastructure | 图片来源：编写组自摄

3.6 建设海绵设施促进水资源集约利用
Build sponge facilities to promote intensified use of water resource

图2-3-18 雨水管断接进入周边海绵设施 / Cut off rainwater pipe and connect it into surrounding sponge facilities | 图片来源：编写组自摄

图2-3-19 高位花坛 / Elevated flowerbed | 图片来源：编写组自摄

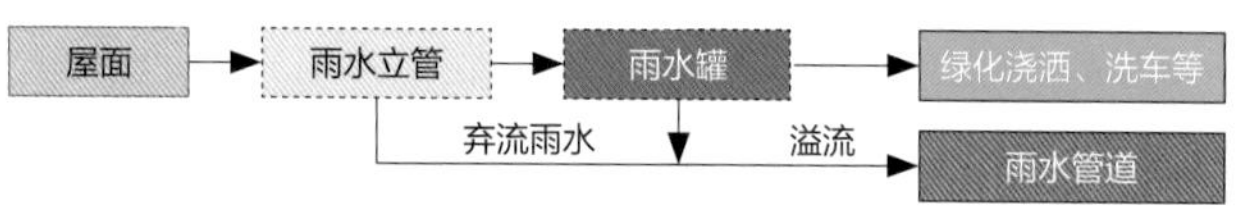

· 雨水立管断接处和雨水罐可通过设计丰富景观的趣味性。
· 雨水罐需放置在平整结实的地面上，底部最好垫高，方便雨水使用。
· At rainwater riser cut-off and rainwater tank, the landscape can be made more interesting by design.
· The rainwater tank shall be put on flat and solid ground, with the bottom elevated to facilitate the use of rainwater.

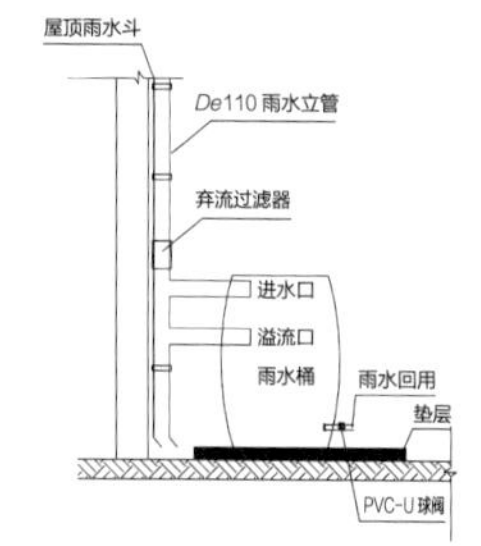

图2-3-20 雨水罐 / Rainwater tank | 图片来源：编写组自绘

图2-3-21 雨水管断接 / Cut off rainwater pipe | 图片来源：编写组自摄

老旧小区海绵化改造一般规定
General stipulations on sponge renovation of old communities

· 遵循先绿色后灰色和先地上后地下、尊重老旧小区本底条件的设计原则。
· 一般采用高位花坛、雨水罐、透水停车场、生态树池、植草沟、下凹式绿地、雨水花园等微小的分散式处理模式。
· Follow the design principles of first green and then gray, first on ground and then underground, and respect the background conditions of old communities.
· Generally, small and distributed facilities such as elevated flowerbed, rainwater tank, permeable parking lot, ecological tree pool, grass ditch, sunken green space and rainwater garden are adopted.

屋面雨水控制和利用
Control and use of rainwater from roof

· 采取雨落管断接或导流设施（雨水管或植草沟）等方式将屋面雨水引入周边绿地内的海绵设施（下凹式绿地、雨水花园等）进行雨水的削污、滞蓄处理。
· 还可通过雨水立管直接断接进入高位花坛或雨水罐进行调蓄，雨水罐内雨水可用于绿化浇洒、洗车等。
· 雨水罐、雨落管设计应注意与景观的协调性。
· Roof rainwater is directed into the sponge facilities (sunken green space, rainwater garden, etc.) in the surrounding green space by means of breaking the rain pipe or diversion facilities (rain water pipe or grass ditch) for treatment to reduce pollution and retention.
· It can also be regulated by directly connecting from rainwater riser into elevated flowerbed or rainwater tank, and the rainwater in the tank can be used to water the plants or wash cars.
· The design of rainwater tank and rain downspout should take into account the coordination with landscape.

道路及广场雨水控制与利用
Control and use of rainwater on roads and squares

- 小区内的人行道、非机动车道、停车场及公共活动广场的更新改造宜采用透水铺装。
- 小区道路排水宜采用生态排水方式，更新改造时应进行适当的竖向优化，使路面坡向绿化带或周边绿地。
- 道路绿化带满足道路雨水径流竖向关系，对有路缘石遮挡的小区道路，可将路缘石做开口更新改造，将雨水引入绿地内。
- Permeable pavement should be used for the renovation of sidewalks, non-motorized driveway, parking lot and public activity square in the community.
- Ecological drainage should be adopted for road drainage in residential areas, and proper vertical optimization should be made in renovation to make the road sloping towards green belt or surrounding green land.
- Road greenbelt shall meet the vertical relationship of road rainwater runoff. For community roads with curb blocks, curbs can be renovated by making openings to introduce rainwater into the green land.

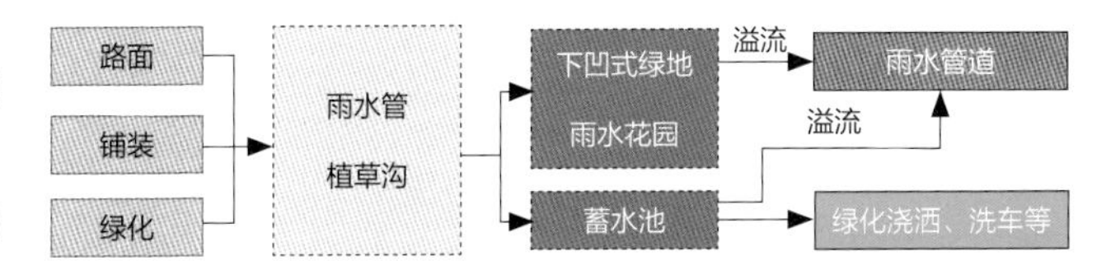

图2-3-22 人行道透水铺装 / Permeable pavement for sidewalks | 图片来源：编写组自摄

图2-3-23 停车场透水铺装 / Permeable pavement for parking lot | 图片来源：编写组自摄

绿地海绵化改造
Sponge renovation for green land

- 绿地是老旧小区海绵化改造的重点，绿地内选用的海绵设施一般包括植草沟、下凹式绿地、雨水花园等。
- 道路两侧绿地设置植草沟，根据小区地形或调整小区竖向，在较低洼处设置下凹式绿地或雨水花园。
- 改造设计时需注意设施中植物的选择和搭配，不仅能发挥海绵设施对雨水的滞蓄作用，还应具备一定的景观效果。
- Green land is the main area of the sponge renovation in old communities. The sponge facilities selected for green land generally include grass ditch, sunken green land, rainwater garden, etc.
- Grass ditches are provided on both sides of roads. According to the landform or by vertical adjustment, sunken green land or rainwater garden can be built in the low-lying area.
- In the renovation design, attention should be paid to the selection and collocation of plants in the facilities, so that they can not only play the role of sponge facilities in rainwater retention, but also produce landscape effect.

图2-3-24 路缘石开口样式 / Cut-out of curbs | 图片来源：编写组自摄

图2-3-25 下凹式绿地 / Sunken green land | 图片来源：编写组自摄

图2-3-26 植草沟 / Grass ditch | 图片来源：编写组自摄、自绘

3.6 建设海绵设施促进水资源集约利用
Build sponge facilities to promote intensified use of water resource

集中雨水收集利用设施
Centralized rainwater collection and utilization facilities

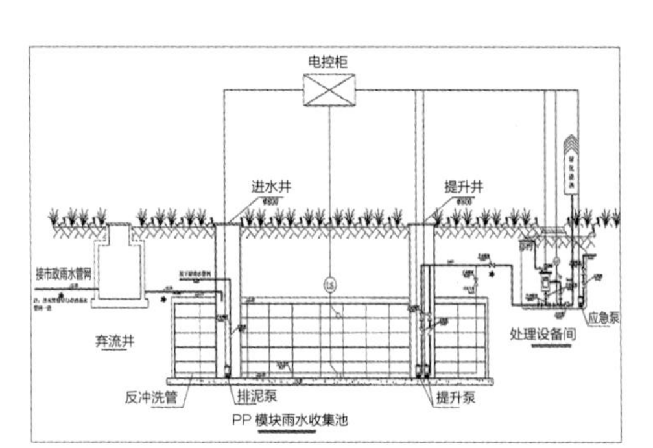

图2-3-27 雨水收集利用系统 / Rainwater collection and utilization system | 图片来源：昆山海创大厦项目海绵专项方案

- 为集约利用水资源，降低小区内涝风险，可以在小区雨水排入市政管网前进入集中雨水收集利用设施，如蓄水池等，经过净化后供小区景观补水及绿化浇洒。
- 小区雨水口设置截污挂篮或者环保雨水口，防止杂物堵塞管道。

- For intensified use of water resources and to reduce the risk of waterlogging in the community, centralized rainwater collection and utilization facilities, such as water pool, can be used to collect rainwater in the community before being discharged into the municipal pipe network. After cleaning, it can be used to replenish water and irrigate the landscape in the community.
- The rainwater port shall be provided with a sewage basket or an environmental protection rainwater port to prevent sundries from blocking the pipeline.

图2-3-28 雨水蓄水箱 / Rainwater storage tank | 图片来源：编写组自摄

图2-3-29 雨水净化设备 / Rainwater cleaning equipment | 图片来源：编写组自摄

污水再生利用
Regeneration and utilization of sewage water

- 有条件的小区可在小区内建设污水处理设施，收集处理小区盥洗废水，处理后用于冲厕、景观喷灌等。
- 污水处理设施尽量利用地下空间，设计要与小区建筑环境相协调。
- 同时设有雨水回用和中水系统时，原水不应混合，出水可在清水池混合。
- 应单独敷设再生水（包括雨水和中水）回用供水管道。再生水供水管道应与生活饮用水管道分开设置，严禁再生水进入生活饮用水给水系统。再生水供水管道不得装设取水龙头，并应采取以下防止误接、误用、误饮的措施：再生水供水管外壁应按设计规定涂色或标识；当设有取水口时，应设锁具或专门开启工具；水池(箱)、阀门、水表、给水栓、取水口均应有明显的“再生水”标识。

- Communities with ready conditions can build sewage treatment facilities, to collect and treat lavatory waste water in the community, and use it after treatment for flushing and landscape spray irrigation.
- Sewage water treatment facilities shall use underground space whenever possible, and the design shall coordinate with the architecture environment of the community.
- When both rainwater recycling and recycled water systems are provided, the raw water should not be mixed, and effluent can be mixed in clean water pool.
- Water supply pipes for reclaimed water (including rainwater and recycled water) shall be laid separately. The supply pipe of reclaimed water should be set up separately from the domestic drinking water pipe. It is strictly prohibited for the reclaimed water to enter the domestic drinking water supply system. The reclaimed water supply pipe shall not be equipped with any faucet, and the following measures shall be taken to prevent misconnection, misuse and improper drinking: the outer wall of the reclaimed water supply pipe shall be painted or marked according to the design provisions; any water intake shall be provided with lock or special opening tool; the water pool (tank), valve, water meter, supply hydrant and water intake shall all be clearly marked with “reclaimed water” .

石家庄天山水榭花都小区中水回用补充景观水、冲厕：
小区住户冲厕、绿化浇洒都用中水，可节约自来水用水量。
小区有两套给水系统，一套自来水，一套中水。
小区内中水处理设备需专人管理，若污水来源不稳定则影响中水供应。

Water is recycled to replenish landscape irrigation and flushing in Tianshan Shuixie Huadu Community in Shijiazhuang:
All residents in the community use recycled water for flushing and plant watering, to save tap water.
The community has two water supply systems, respectively for tap water and recycling water.
The recycling water treatment equipment is managed by dedicated personnel, and instable source of sewage water will affect the supply of recycling water.

Construction Guidance for
Renovation of Old Communities in Jiangsu

江苏老旧小区改造建设导引

4

改善交通及停车设施
Improve Traffic and Parking Facilities

- 4.1 **交通序化与道路设施更新维护**
  Good traffic order and updating and maintenance of road facilities
- 4.2 **满足日益增长的停车需求**
  Meet the growing demand for parking
- 4.3 **设置电动自行车充电设施**
  Set up charge facilities for electric bicycles
- 4.4 **提升小区人行环境品质**
  Upgrade community pedestrian environment quality
- 4.5 **出入口空间优化的几种方式**
  Methods to optimize access space
- 4.6 **完善道路稳静化设施**
  Complete facilities to stabilize and quiet roads
- 4.7 **设置或预留电动汽车充电设施**
  Set up or reserve charge facilities for electric automobiles
- 4.8 **建设立体停车设施**
  Build 3D parking facilities
- 4.9 **运用智慧手段实现小区内外停车设施共享**
  Adopt intelligent means to share parking facilities inside and outside communities

4.1 交通序化与道路设施更新维护
Good traffic order and updating and maintenance of road facilities

梳理主次道路系统，优化交通流线
Sort out main and secondary road systems, and optimize traffic flow lines

- 结合小区空间结构，梳理优化主、次道路系统，通过进出分离、单向组织等方式，序化小区交通。
- 优化各级道路红线宽度和路幅分配，保持车行、人行交通顺畅、安全，以及满足消防、救护等车辆通行要求。
- 扣除停车空间，双向通行的道路宽度不宜小于5.5m，单向通行的道路宽度不宜小于3m。

- Main and secondary road systems are sorted out in conjunction with the space structure of the community, to put the community traffic in order by separating inflow and outflow and one-way traffic organization.
- Optimize the red line width and lane distribution for roads of different grades, to maintain smooth and safe vehicle and pedestrian traffic, and also meet the passage of fire engines and ambulance.
- After deducting the parking space, the road width should be no less than 5.5m for two-way traffic and no less than 3m for one-way traffic.

单向机动车道（消防通道）>4.5m　停车带 2.5m　人行道 >1.5m

小区主路单向通行断面（6~8.5m）

图2-4-1 小区道路断面优化建议 / Suggestions on optimization of community road sections | 图片来源：编写组自绘

图2-4-2 小区道路拓宽使交通更加通畅 / Widen the road in community to make traffic more smooth | 图片来源：南京金尧山庄老旧小区整治材料

修补破损路面
Repair damaged road surface

- 对于只是面层龟裂、坑槽、沉陷，道路基层、垫层质量较好的道路，可对其进行局部面层铲除，用原面层材料重新进行铺设，或者采用新材质对面层重新进行铺整。
- 对于破损严重的道路，应重新进行铺整，其面层、基层、垫层构造应根据道路性质的荷载要求进行设计。

- For roads with only surface cracks, pits and subsidence, and with good subgrade and bedding quality, the pavement can be removed partly, and be paved again with the same materials as original, or pave the surface again with new materials.
- Roads seriously damaged should be repaired and paved again, and the surface, subgrade and bedding shall be designed according to the load requirements based on the application of road.

小区路面材质选择
Selection of road surface materials

- 小区道路宜采用柔性路面。
- 宅间路可采用刚性路面。
- 人行道宜采用透水性较好的砌块路面。

- Flexible surface should be used for community roads.
- Rigid surface can be used for roads between houses.
- Well-permeable brick pavement should be used for sidewalks.

图2-4-3 小区道路宜采用沥青混凝土路面 / Bituminous concrete surface should be used for community roads | 图片来源：编写组自摄

图2-4-4 宅间路宜采用刚性路面 / Rigid surface should be used for roads between houses | 图片来源：编写组自摄

图2-4-5 步行道宜采用透水砖路面 / Water permeable surface should be used for sidewalks | 图片来源：编写组自摄

4.2 满足日益增长的停车需求
Meet the growing demand for parking

图2-4-6 结合小区道路布设停车泊位 / Arrange parking positions in combination of roads in community | 图片来源：编写组自摄

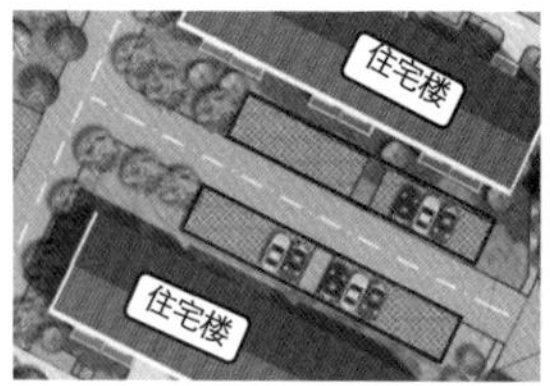

图2-4-7 利用住宅背向院落布设停车泊位 / Arrange parking positions in courtyards on the back of houses | 图片来源：左图：编写组自绘；右图：昆山中华北村小区改造规划

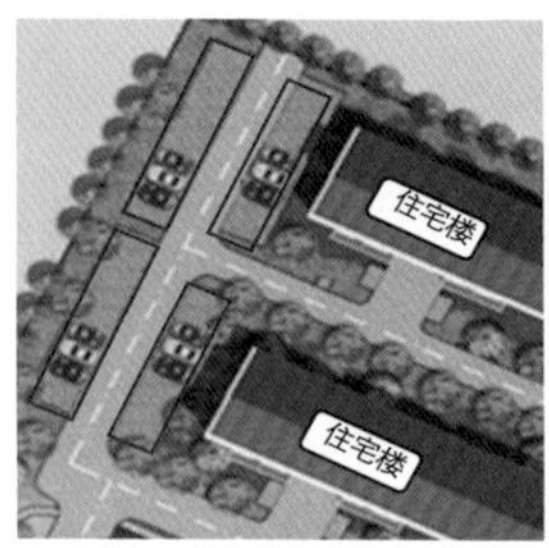

图2-4-8 利用边角零星用地布设停车泊位 / Arrange parking positions in marginal land | 图片来源：左图：编写组自绘；右图：昆山中华北村小区改造规划

合理确定停车泊位配置规模
Rationally determine parking position allocation and size

· 原则上不低于改造前的现状停车泊位供给规模。
· 处理好停车设施与景观绿化之间的协调关系，适当考虑未来停车需求。
· In principle, no less than the status quo size of parking positions before renovation.
· Ensure harmonious relationship between parking facilities and landscape plantation, with due consideration on the future parking needs as appropriate.

江苏老旧小区改造案例中停车泊位配置情况
Parking position allocation in old community renovation cases in Jiangsu

改造小区名称 Name of renovated community	小区户数 No. of households	规划停车泊位 Planned parking positions	户均泊位 Position per household
南京金尧山庄 Jinyao Villa, Nanjing	508	290	0.57 个 / 户 0.57/household
盐城南苑小区 Nanyuan Quarter, Yancheng	832	340	0.41 个 / 户 0.41/household
昆山中华园北村 Zhonghua Garden North, Kunshan	980	317	0.32 个 / 户 0.32/household
昆山枫景苑 A 区 Fengjingyuan Zone A, Kunshan	1380	615	0.45 个 / 户 0.45/household
昆山琼花新村 Qionghua New Village, Kunshan	1046	284	0.27 个 / 户 0.27/household

增补机动车停车设施的方式
Ways to increase motorized vehicle parking facilities

· 结合小区空间条件，因地制宜采用集中和分散、地面和立体相结合的方式布置机动车停车泊位。
· 结合小区道路沿线空间采用垂直、斜列或平行方式排列泊位。
· 利用住宅背向院落布设停车泊位。
· 利用边角零星用地布设停车泊位。
· Arrange parking positions for motorized vehicles by combining centralized and distributed parking and ground and elevated parking according to the space conditions of the community.
· Arrange parking positions in vertical, oblique or parallel patterns along roads in the community.
· Arrange parking positions in courtyards on the back of houses.
· Arrange parking positions in marginal land.

停车场地设计
Design of parking lot

- 场地铺装材质：以生态植草格和沥青为主，可以彩色区分其他铺装。
- 停车标识设计：完善停车地面标识、标线、编号等要素，规范停放区域。
- 停车遮蔽设施：通过树木、棚架等设施提供遮阳、避雨功能。
- Ground paving material: ecological grass grid and asphalt as the main, can be distinguished with different colors.
- Parking sign design: complete the parking ground sign, line mark, numbering and other elements, and standardize the parking area.
- Parking shelter facilities: provide shade and rain shelter with trees, shed and other facilities.

停车设施智能化管理
Intelligent management of parking facilities

- 包括自动道闸、感应卡读感器、感应卡、语音提示等，一般设置在小区出入口，与行人出入口相分离。
- 业主车辆：采用视频车牌识别，车辆进入小区，道闸自动开启，快捷通畅。
- 临时车辆管理：访客车辆经保安登记确认后给予临时授权通行，按小时计费。
- Including automatic lane gate, induction card reader, induction cards and voice prompt, etc., which are generally set at the entrance of the community and separated from the entrances of pedestrians.
- Owner cars: video license plate recognition shall be used, when a car enters the community, the gate automatically opens, quick and unimpeded.
- Temporary car management: the visitor cars will be authorized to pass after security registration and confirmation and be charged by hours.

图2-4-9 车辆智能化管理系统 / Vehicle intelligent management system | 图片来源：编写组自摄

4.3 设置电动自行车充电设施
Set up charge facilities for electric bicycles

完善非机动车停车设施
Complete parking facilities for non-motorized vehicles

- 非机动车停车设施布局以分散、地面方式为主，方便居民停放。
- 小区已有地下（半地下）非机动车停车设施的，可结合住宅楼栋入口设置临时停放车位。
- 小区没有地下（半地下）非机动车停车设施的，则结合住宅院落、宅间路、宅旁空间划分非机动车停车泊位，辅以遮阳棚架。
- 非机动车停车区可通过铺装、划线、固定装置和相关标识予以区分，与周边道路和建筑相协调。
- 非机动车车棚不得影响周边住宅通风采光，宜采用轻型材质建造，造型轻巧，色彩与周边环境协调，并配置充电插座、充气筒等设施。

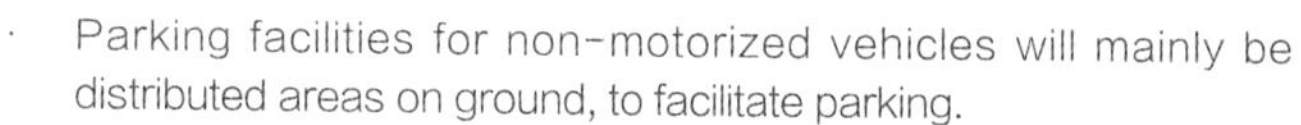

- Parking facilities for non-motorized vehicles will mainly be distributed areas on ground, to facilitate parking.
- In communities with underground (semi-underground) parking for non-motorized vehicles, temporary parking positions can be set up at residence building entrance.
- In communities with no underground (semi-underground) parking for non-motorized vehicles, parking ground for non-motorized vehicles shall be planned by courtyards, road between houses and space by houses, and sun-shading shed will be built.
- Parking areas for non-motorized vehicles can be distinguished by pavement, marking, fixed devices and relevant signs, to be harmonic with roads and buildings nearby.
- Sheds for non-motorized vehicles must not affect the ventilation and lighting of nearby houses, they should be made of light materials, with colors harmonic with surrounding environment, and be provided with facilities such as charge sockets and inflator pumps.

图2-4-10 非机动车车棚 / Shed for non-motorized vehicles | 图片来源：编写组自摄

分片集中设置电动车充电设施
Set up centralized charge facilities for electric automobiles at different areas

· 可与非机动车车棚相结合，配套自动断电、故障报警等功能，确保使用安全。

· They can be combined with sheds of non-motorized vehicles, and provided with functions of automatic power cut-off and fault alarm, to ensure safety in use.

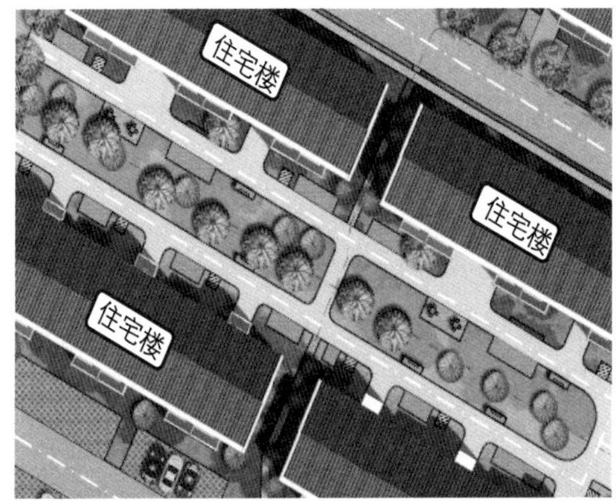

图2-4-11 结合楼栋入口布设非机动车停车设施 / Parking facilities for non-motorized vehicles close to building entrance | 图片来源：上图：编写组自绘；下图：编写组自摄

图2-4-12 结合小块空地布设非机动车停车设施 / Parking facilities for non-motorized vehicles in small empty land | 图片来源：昆山中华北村小区改造规划、南京金尧山庄老旧小区整治材料

4.4 提升小区人行环境品质
Upgrade community pedestrian environment quality

完善小区步道系统
Complete the community footpath system

- 步道系统：包括人行道、独立休闲步道等，步道系统应连接小区入口、住宅、公建、公共活动空间、公交车站等功能设施。
- 小区主路应设置人行道：结合空间条件可以采取单侧或双侧设置方式，亦可通过划线或色彩、标志等方式区分，人行道宽度不宜小于1.5m。
- 独立休闲步道：可利用小区主路沿线空间、宅旁空间等进行设置，相互连通，方便小区居民日常步行活动和出行。
- 步道人性化设施：步道宜采用透水砖、透水沥青、透水木塑板等材料，兼顾舒适性、生态性和景观性；结合景观塑造，局部可增设遮阳设施，形式宜轻巧、透空，具有观赏性和艺术性。

- Footpath system: including sidewalks, independent leisure footpaths, etc. The footpath system shall connect the community entrance, residential buildings, public buildings, public activity space, bus station and other functional facilities.
- The main road in the community shall be provided with sidewalk: according to the space conditions, it can be arranged either on one side or on both sides, and can also be distinguished by lines, colors, signs and other ways. The width of the sidewalk should be no less than 1.5m.
- Independent leisure footpath: can be set up in the space along the community main road and the space by houses and connected to each other, to facilitate the daily walking activities and travel of residents in the community.
- Humanized facilities for footpath: the footpath should be made of permeable bricks, permeable asphalt, permeable wood plastic board and other materials, giving consideration to comfort, ecology and landscape. Combined with landscape shaping, local shading facilities can be added, in a light and hollow form, with ornamental and artistic effect.

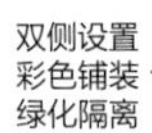

图2-4-13 小区主路人行道设置方式 / Sidewalk setup form along community main road | 图片来源：南京尧林仙居老旧小区整治材料

图2-4-14 步道沿线布设座椅、遮阳廊架等人性化设施 / Humanized facilities such as seats and sunshade corridor are arranged along the footpath | 图片来源：编写组自摄

▶ 因地制宜实现人车分流
Separate people and vehicle flows according to local conditions

· 结合小区空间条件，采取调整道路功能、优化路网结构、组织单向交通等方式，实现小区整体或组团的人车分流，尽可能减少人车冲突与矛盾。

· Separate people and vehicle flows in the whole community or in groups according to the space conditions by adjusting road functions, optimizing road network structure and organizing one-way traffic, to minimize conflicts between people and vehicles.

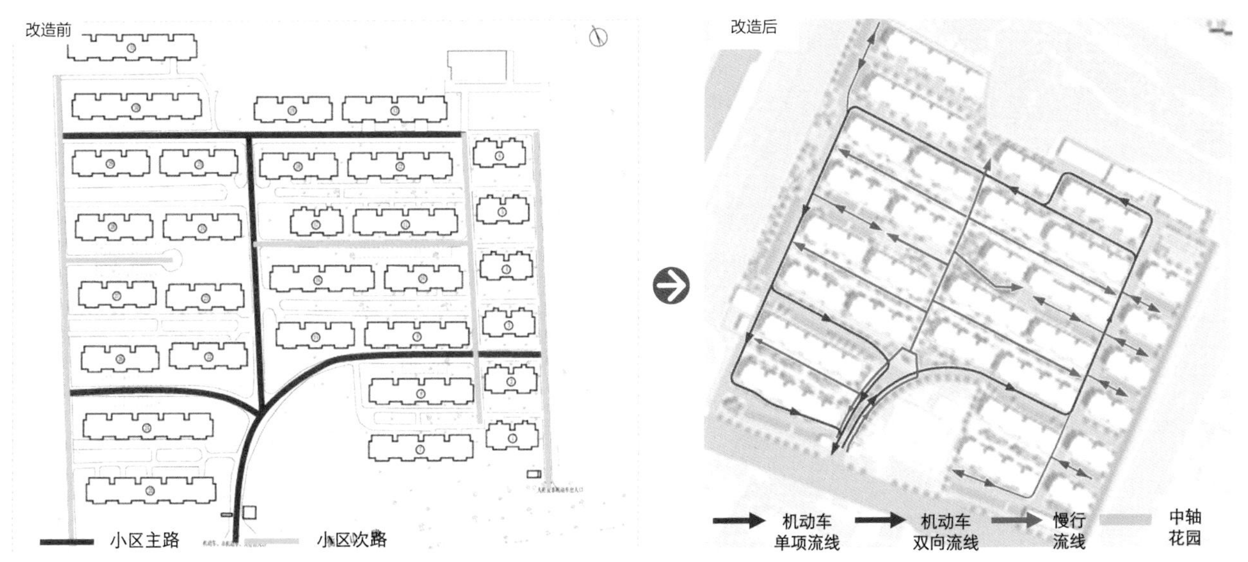

图2-4-15 昆山中华园北村通过优化交通组织实现人车分流 / Zhonghua Garden North in Kunshan realizes separation of people and vehicles by optimizing traffic organization | 图片来源：昆山中华北村小区改造规划

图2-4-16 昆山中华园北村中轴休闲步道改造效果 / Renovation effect of central axis leisure footpath in Zhonghua Garden North in Kunshan | 图片来源：昆山中华北村小区改造规划

4.5 出入口空间优化的几种方式
Methods to optimize access space

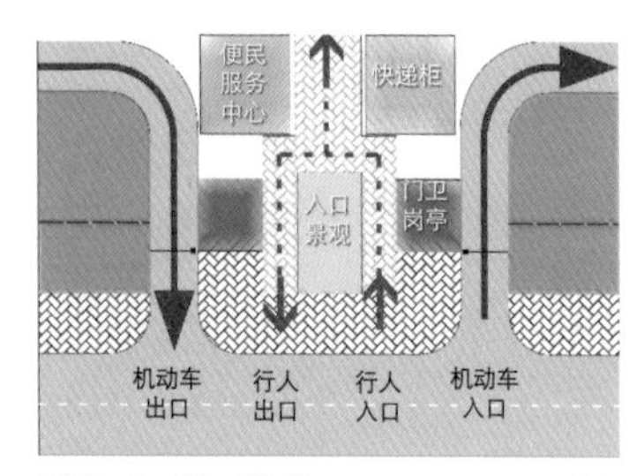

图2-4-17 模式一 / Mode I | 图片来源：编写组自绘

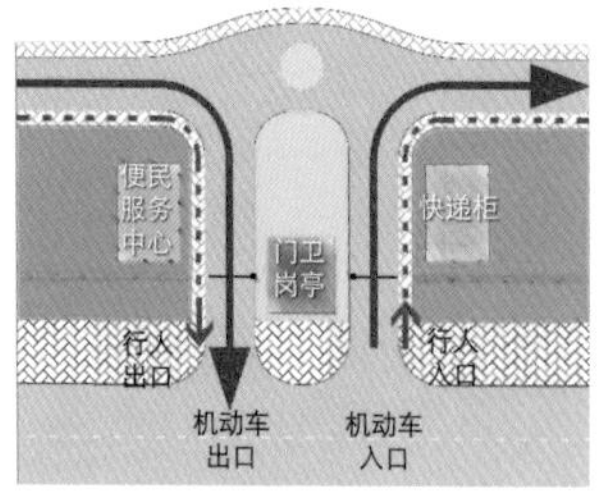

图2-4-18 模式二 / Mode II | 图片来源：编写组自绘

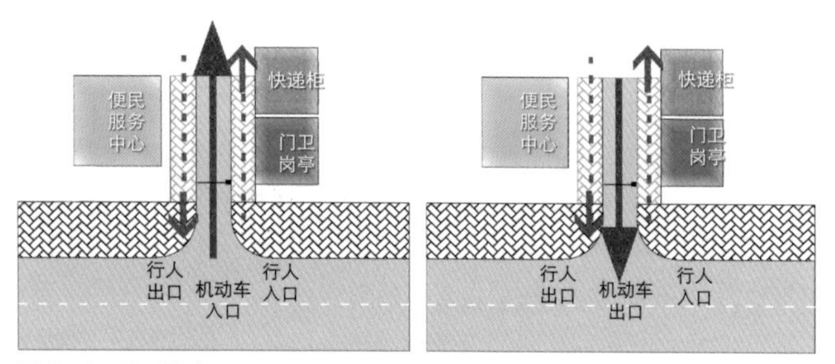

图2-4-19 模式三 / Mode III | 图片来源：编写组自绘

▶ 小区主要出入口宜通过绿岛形式实现进出分离
Main entrance and exit should be separated by green island

- 模式一：人行居中、车行两侧。
- 模式二：人车同向、进出分离。
- 模式三：单向进出、人车分离。
- Mode I: people in the middle and vehicles on both sides.
- Mode II: people and vehicles moving in the same direction, with entering and exiting flows separated.
- Mode III: separated people and vehicles flows, with one-way entering and exiting.

▶ 合理布局出入口各类设施
Rationally arrange various access facilities

- 入口设施包括门卫、岗亭、快递收发、便民服务中心、宣传公告栏等，应整体设计，做到流线合理、有序。
- Entrance facilities include guard box, express delivery, convenience service center and publicity bulletin, and they should be designed as a whole for rational and orderly flow lines.

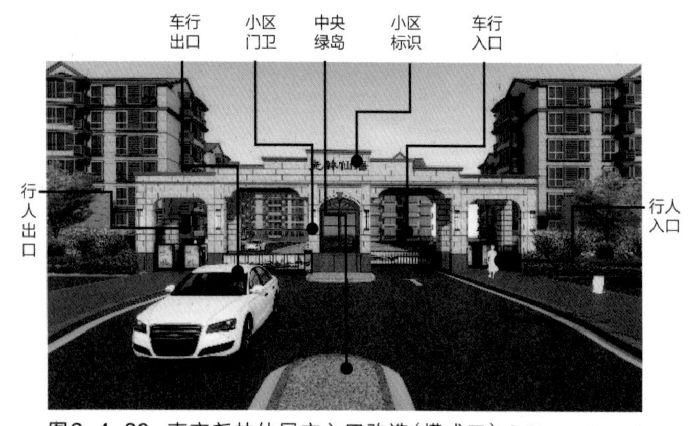

图2-4-20 南京尧林仙居主入口改造（模式二）/ Renovation of main entrance of Yaolin Xianju in Nanjing (Mode II) | 图片来源：南京尧林仙居小区改造材料

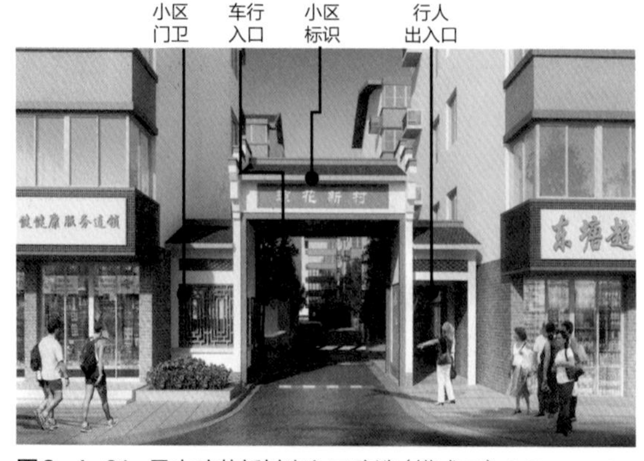

图2-4-21 昆山琼花新村出入口改造（模式三）/ Renovation of entrance and exit of Qionghua New Village in Kunshan (Model III) | 图片来源：昆山琼花新村小区改造材料

4.6 完善道路稳静化设施
Complete facilities to stabilize and quiet roads

完善交通标志标线
Complete traffic marking lines

- 通过道路标线和交通标识，引导机动车、非机动车和行人各行其道，减少不同流线的相互冲突。

- Guide the motorized vehicles, non-motorized vehicles and pedestrians on the right route with road marking lines and traffic signs, to reduce conflict of different flow lines.

图2-4-22 施划机动车导向箭头、泊位边线、人行道横道等交通标线 / Traffic mark lines such as motorized vehicle guide arrows, parking position edge lines and pedestrian crossing lines | 图片来源：编写组自摄

图2-4-23 设置限速、禁止鸣笛等交通标志 / Set up traffic signs such as speed limit and no horn | 图片来源：编写组自摄

设置减速带等稳静化设施
Set up stabilizing and quieting facilities such as speed bumps

- 小区主要出入口和人车交织的地点应设置减速带，在车辆视距受限的转角处设置凸面转角反光镜，保障小区道路的行人步行安全。

- Speed bumps should be set at the main entrances and exits and at intersection of vehicles and people, and convex turning mirrors should be set at corners with limited visual distance of vehicles, to ensure pedestrian safety on road in the community.

图2-4-24 小区进出口处设置减速带 / Speed bumps at community entrance and exit | 图片来源：编写组自摄

4.7 设置或预留电动汽车充电设施
Set up or reserve charge facilities for electric automobiles

提供电动汽车充电设施场地
Provide ground for electric automobile charge facilities

· 结合小区实际情况和未来发展需求，配置或预留电动汽车充电桩，应综合考虑供电线缆铺设、配电箱安装、消防安全、后期维护等多样因素，满足相关安全要求。

· According to the actual situation and the future development needs of the community, electric automobile charging piles should be provided or reserved, and various factors such as power cable laying, distribution box installation, fire safety, and subsequent maintenance shall be comprehensively considered to meet the relevant safety requirements.

充电设施建设与管理
Construction and management of charge facilities

· 利用小区的公共车位建设充电桩。
· 引入充电桩企业作为第三方进行小区充电桩建设和运营，并提供后续设备维护和管理服务。
· 物业公司负责充电设施的日常运行、设备看护和车位管理工作。

· Build charge piles at public parking positions in the community
· Introduce charge pile enterprises to build and operate charge piles in the community as a third party, and provide follow-up equipment maintenance and management services.
· The property management company is responsible for the daily operation of charge facilities, equipment care and parking space management.

图2-4-25 利用公共泊位设置电动汽车充电桩 / Set up charge piles for electric automobiles at public parking positions | 图片来源：南京金尧山庄老旧小区整治材料

4.8 建设立体停车设施
Build 3D parking facilities

▶ 挖掘小区存量空间建设立体停车设施
Tap inventory space in the community to build 3D parking facilities

· 一般宜选择在小区边缘、对小区交通干扰少、景观影响小的地段，建设立体停车设施。

· Generally, it is appropriate to build 3D parking facilities at the edge of the community, where there is less disturbance to traffic and less impact to landscape.

▶ 利用屋顶、地下空间设置停车设施
Use roof and underground space to set up parking facilities

· 鼓励空间复合利用，利用公共建筑屋顶增设停车设施；结合小区公共空间改造建设半地下停车场。

· Encourage the diversified utilization of space and add parking facilities on the roof of public buildings; Renovate and build semi-underground parking lot in community public space.

图2-4-26 机械立体停车 / 3D mechanical parking building | 图片来源：编写组自摄

4.9 运用智慧手段实现小区内外停车设施共享
Adopt intelligent means to share parking facilities inside and outside communities

利用周边资源，缓解老旧小区的停车矛盾
Mitigate parking problems in old communities by using surrounding resources

- 充分利用小区周边资源补充停车设施，如利用周边非交通性道路或支路设置夜间临时停车位。
- 利用周边学校操场、公园绿地等建设地下公共停车场。

- Make full use of the surrounding resources to supplement parking facilities, such as the use of the surrounding non-traffic roads or branches as temporary parking spaces at night.
- Use the playground of the surrounding schools, parks and green land to build underground public parking lots.

图2-4-27 利用周边支路设置路内停车泊位 / Use surrounding branch roads as road parking space | 图片来源：编写组自摄

图2-4-28 利用学校操场建设地下公共停车场，缓解停车供需矛盾 / Build underground public parking lots under school playground to mitigate the contradiction between parking supply and demand | 图片来源：编写组自摄

运用互联网技术建立停车设施共享机制
Use Internet technology to establish parking facilities sharing mechanism

- 鼓励住区周边机关企事业单位、公共建筑、学校等，向周边居民错时开放内部停车场。
- 依托社区建立停车信息共享平台，或加入城市停车信息共享平台，实现泊位错时共享，缓解停车难的问题。

- Encourage enterprises and institutions, public buildings and schools in the surrounding areas to open internal parking lots to the surrounding residents in off-time.
- Establish parking information sharing platform in the community, or join the urban parking information sharing platform, to share parking space at different times, to alleviate the problem of difficult parking.

- 结合共享App服务，智慧化解决空置车位问题。
- 共享车位宜通过颜色重点区分。
- 增加新能源车充电桩。

- Make use of idle parking positions in an intelligent way with sharing App service.
- Sharing positions should be marked with different colors.
- Add charge piles for new energy vehicles.

图2-4-29 共享停车App / Parking share App | 图片来源：编写组自摄

Construction Guidance for
Renovation of Old Communities in
Jiangsu

江苏老旧小区改造建设导引

5

保持小区环境整洁卫生
Keep the Neighborhood Clean and Tidy

5.1 清理小区脏乱环境
Clean the neighborhood environment

清理垃圾杂物
Clean garbage and sundries

- 小区物业负责及时清理小区道路、公共空间、楼宇间和楼道内的垃圾。

- The property management of the community shall timely clean up the garbage in the roads, public spaces, buildings and stairways in the community.

清理乱贴乱画
Clean graffiti

- 小区物业负责及时清理乱贴乱画、乱写乱刻现象，保证小区环境干净整洁。
- 结合小区出入口、公共活动空间设置公告栏、宣传栏，集中发布信息，公告栏样式应与小区整体环境相协调。

- The property management of the community shall promptly clean the graffiti and scribbles to ensure that the residential environment is clean and tidy.
- Bulletin boards and publicity boards shall be set up at the community entrances and in the public activity space, to release information in a centralized manner. The style of bulletin boards shall be harmonic with the overall environment of the community.

图2-5-1 结合小区入口设置公告宣传栏 / Bulletin board set up at community entrance | 图片来源：编写组自摄

图2-5-2 结合公共区域设置公告宣传栏 / Bulletin board set up in public space | 图片来源：编写组自摄

5.2 加强小区日常环卫保洁工作
Strengthen routine cleaning for neighborhood environment

根据人口规模配置保洁人员
Allocate cleaning personnel according to population size

· 物业应按照小区人口规模配备足够数量的保洁人员，建立健全并落实小区清扫保洁、分片包干、巡查监督等管理制度，做到垃圾日产日清。

· The property management shall allocate enough cleaning personnel according to the population size of the community, establish, complete and implement the management rules such as community cleaning and maintenance, contracting by zones, tour inspection and supervision, etc., to ensure that garbage is cleaned up on the same day it is produced.

配备专业垃圾收运车辆
Provide professional garbage collection and transport vehicles

· 物业应配备专业垃圾清运车辆，实行密闭化运输，避免垃圾清运中渗滤液所带来的二次污染。

· The property management shall be equipped with professional garbage collection and transport vehicles for enclosed transport, to avoid secondary pollution caused by leachate in garbage collection and transport.

图2-5-3 专业车辆保障垃圾密闭运输 / Professional vehicle ensure enclosed transport of garbage | 图片来源：上图：编写组自摄；下图：江苏省城市环境综合整治技术指南

5.3 设置生活垃圾分类投放设施
Set up facilities for classification of living wastes

图2-5-4《江苏省城市居民生活垃圾分类投放与收运设施配置指南》(试行)将居民生活垃圾分为四类 / The domestic wastes of residents are classified into four categories according to the *Guide to classified disposal of domestic wastes from urban residents and allocation of collection and transport facilities in Jiangsu Province* (for trial implementation) | 图片来源：昆山市亭林城市管理办事处琼花新村小区改造项目成果

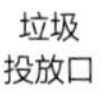

图2-5-5 垃圾分类收集房 / Classified garbage collection room | 图片来源：昆山市亭林城市管理办事处琼花新村小区改造项目成果

- 垃圾分类收集房配备有害垃圾、可回收物、厨余垃圾、其他垃圾四类垃圾投放口。
- 收集房配有洗手池。
- 外墙有电子宣传显示屏和垃圾分类宣传图画。
- The garbage classified collection room shall be provided with four garbage ports: harmful garbage, recyclables, kitchen waste and other garbage.
- The collection room shall be provided with a wash basin.
- Electronic publicity display screen and waste classification publicity pictures shall be provided on the exterior wall.

生活垃圾投放方式
Ways of discarding domestic wastes

- 采用“四分类”法，将居民生活垃圾按照可回收物、有害垃圾、厨余垃圾、其他垃圾进行分类。
- 提倡撤桶并点、设置垃圾箱间（房），推行生活垃圾定时定点分类收集。
- Domestic wastes from residents are classified into four categories: recyclables, harmful garbage, kitchen waste and other garbage.
- It is advocated to remove some garbage bins to put them together, set the garbage bin rooms, and collect domestic wastes at fixed time and places after classification.

集中设置垃圾分类收集间（房）
Set up rooms for classified collection of garbage

- 生活垃圾分类收集容器间（房）应选择利于清运的地点设置，宜设置于小区出入口、主要道路旁及对交通、居民生活、景观效果影响较小的地段，服务户数一般不超过1000户，服务半径不宜超过300m。
- 分类收集容器间（房）配备分类垃圾桶、分类指引说明、洗手池、排水口等配套设施，并安排专人进行管理和垃圾分类督导。
- 分类收集容器间（房）应根据场地空间，选择合适的布置形式，场地应做硬化和防水处理，完善排水、消毒、消防、空气净化等设施。
- The rooms for classified collecting containers of domestic wastes should be located where transport is convenient, such as community entrances and exits, and by the main roads, with little impact on traffic, residents' life and landscape effect. The number of households served of one room should not exceed 1000 in general, with service radius not exceeding 300m.
- A room for classified collecting containers shall be provided with classification garbage bins, classification instructions, wash basins, drainage outlets and other supporting facilities, and personnel shall be assigned to manage and supervise the classification of garbage.
- A room for classified collecting containers shall be in an appropriate layout according to the site space, the ground shall be hardened and water-proof, with complete drainage, disinfection, fire protection, and air cleaning facilities.

分散设置垃圾分类收集点（容器）
Scattered spots (containers) for classified collection of garbage

- 暂无条件采用垃圾收集容器间（房）集中投放生活垃圾的小区，可对现有的垃圾投放量和户外空间进行重新评估，采用分散设置的垃圾分类收集点（容器）进行垃圾分类收集。分散设置的垃圾分类收集点（容器）宜逐步减少。
- 宜按单元或楼栋在适宜位置分别设置收集厨余垃圾、其他垃圾的垃圾容器。高层住宅按照每栋，小高层和多层住宅按照每1~2栋，设置一组可回收物收集点（容器）。有害垃圾收集容器宜在小区各主要出入口设置，并应采用满足有害垃圾处理要求、具备分隔作用的专用收集箱。

- In communities tentatively unable to set up a garbage collection room for centralized disposal of domestic wastes, the current garbage disposal amount and outdoor space can be evaluated again, and scattered spots (containers) for classified collection of garbage can be used for classified collection of garbage. Scattered spots (containers) for classified collection of garbage should be gradually reduced.
- Garbage containers for separately collecting kitchen waste and other garbage should be provided at appropriate locations in units or buildings. A set of recyclable collection containers shall be provided for each high-rise residential building, or for every 1~2 small high-rise and multi-storey residential buildings. Harmful garbage collection containers should be set up at the main entrances and exits of the community, and special collection boxes with separation function should be used to meet the requirements of harmful garbage disposal.

废旧衣物收集箱
Waste clothes collection boxes

- 每个小区结合主要出入口、主要通道设置废旧衣物收集箱，服务半径不宜超过300m。设施应密闭、防雨、防潮、防腐、阻燃，并具备专人开启功能。

- Each community shall be provided with waste clothes collection boxes at main entrances and exits, with service radius not exceeding 300m. The facilities shall be enclosed, rainproof, moisture-proof, anticorrosive and flame-retardant, and have a dedicated person to open them.

图2-5-6 小区出入口设废旧衣物收集箱 / Waste clothes collection box at community entrance or exit | 图片来源：编写组自摄

5.3 设置生活垃圾分类投放设施
Set up facilities for classification of living wastes

设置建筑垃圾、大件垃圾堆放点
Heaping spot for construction rubbish and large garbage shall be set up

· 结合小区规模，设置相对集中的建筑垃圾堆放点，可结合小区次要出入口设置，便于车辆通行且不影响小区景观。
· 建筑垃圾、大件垃圾堆放点应与住宅保持卫生、消防安全距离，并应设置围挡，防止扬尘。

· According to the size of community, relatively centralized construction rubbish heaping spot shall be provided, and it can be located at secondary entrance or exit, facilitating vehicle passage and not affecting the landscape of the community.
· Heaping spot for construction rubbish and large garbage shall keep a hygiene and fire protection safety distance from resident houses, and fence shall be set up to prevent flying dust.

上海彭浦镇佳宁花园小区建筑垃圾厢房改造：
· 包含大件建筑垃圾堆放点、小件干垃圾堆放点。
· 内置喷淋系统，降低扬尘。
· 外墙绘制垃圾分类宣传画。
Renovation of wing-room for construction rubbish in Jianing Garden of Pengpu Town, Shanghai:
· Including large construction rubbish heaping and small dry rubbish heaping.
· With internal spray system to reduce flying dust.
· Publicity picture on garbage classification plotted on exterior wall.

开展垃圾分类宣传活动
Carry out publicity on garbage classification

· 物业应积极宣传推动垃圾分类行动，可以采取积分兑换制度、垃圾分类宣传活动、社区游戏等方式，鼓励居民实施垃圾分类。

· The property management shall actively promote garbage classification, to encourage residents to implement garbage classification in forms such as point exchange system, garbage classification publicity activities, community games, etc.

图2-5-7 垃圾分类宣传 / Publicity on garbage classification |
图片来源：编写组自摄

5.4 运用智能设备提高垃圾分类收运效率
Use intelligent equipment to better classify wastes

采用智能投放设备
Use intelligent equipment for garbage

· 有条件的小区，可采用具有智能辨识类别功能、自动计量的智能垃圾分类投放设备，提高垃圾分类投放管理效率。

· If conditions permit, intelligent garbage sorting equipment with intelligent identification function and automatic measurement can be used, to improve the efficiency of garbage classification and disposal management.

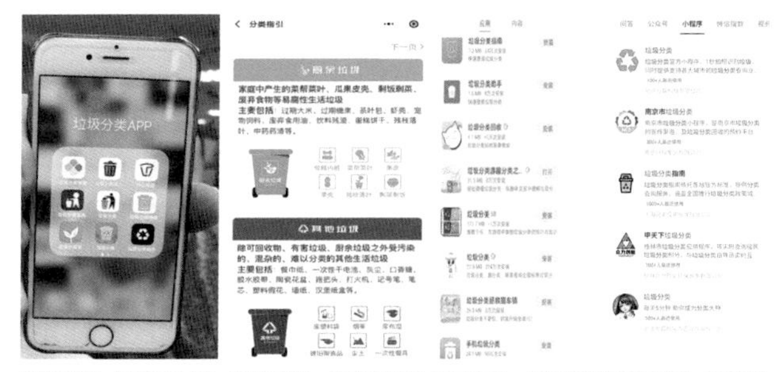

图片来源：南京市垃圾分类小城区手机截图 图片来源：南京市垃圾分类小城区手机截图-2 图片来源：手机截图 图片来源：手机截图-2

图2-5-8 垃圾分类App / Garbage classification App

实现小区垃圾分类智能化、数据化管理
Realize intelligent and data management of garbage classification

· 通过智能垃圾分类系统，管理部门可以掌握小区垃圾分类的工作情况，适时改进工作。

· With the intelligent garbage classification system, management department can know the work situation of garbage classification in the community and timely improve the work.

图2-5-9 智能垃圾分类设备 / Intelligent garbage sorting equipment | 图片来源：编写组自摄

Construction Guidance for

Renovation of Old Communities in Jiangsu

江苏老旧小区改造建设导引

6

方便居民日常生活
Make Daily Life More Convenient

- 6.1 老旧小区5分钟服务设施配置内容
  Configuration contents for 5 minutes service facilities in old communities
- 6.2 合理设置快递投放位置
  Rationally arrange locations for express delivery
- 6.3 设置老年助餐点
  Set up meal service places for the elderly
- 6.4 统一布置室外晾晒设施
  Arrange unified outdoor air drying facilities
- 6.5 提供方便可寻的公厕
  Provide convenient public toilets
- 6.6 为居民和来访者提供清晰的指引标识
  Provide clear guide signs for residents and visitors
- 6.7 有条件的小区可设置老年日间照料中心
  Set up day care centers for the elderly in communities with ready conditions
- 6.8 适当预留场地布设临时便民设施
  Make appropriate reservation for temporary facilities to facilitate people
- 6.9 挖掘周边资源，进一步完善社区服务
  Tap surrounding resources to further consummate community services
- 6.10 提供服务便利的商业网点
  Provide convenient store and service spots
- 6.11 提供就近的社区卫生服务
  Provide community health service nearby
- 6.12 提供5分钟可达的公交站点
  Arrange bus stops accessible in 5 minutes

6.1 老旧小区 5 分钟服务设施配置内容
Configuration contents for 5 minutes service facilities in old communities

老旧小区 5 分钟服务设施配置内容
Configuration contents for 5 minutes service facilities in old communities

老旧小区5分钟服务设施配置内容

事项		应配建项目	按需配建项目	设施布局要求
社区服务	社区用房	物业管理	小区议事厅、党群服务站	宜布置在小区内部，可结合其他公共服务设施联合建设
	幼托设施		幼儿园	可结合小区周边统筹设置
	老年服务	老年助餐点	日间照料中心	可结合小区周边统筹设置
	卫生服务站		社区卫生服务站	可结合小区周边统筹设置
	文化活动站		图书阅览室、多功能活动室（棋牌、亲子）	可结合小区周边统筹设置
健身活动设施		室外健身器械	环形健身步道	宜结合小区广场绿地设置
		儿童活动场地		宜结合小区广场绿地设置
		老年活动场地		宜结合小区广场绿地设置
便民服务		公厕、快递设施、再生资源回收点		宜结合小区出入口、主要通道设置
		室外晾晒设施		可利用住宅院落空间设置
商业网点		便利店、药店、洗衣店、理发店	菜市场、宠物店、餐饮店、银行网点	可结合小区周边统筹设置
公交设施		公交站点 / 轨道站点		结合小区周边街道设置

Configuration contents for 5 minutes service facilities in old communities

Item		Compulsory item	Item as needed	Layout requirements
Community services	Community management room	Property management	Community meeting hall Party and mass service station	Located inside the community, can be built jointly with other public service facilities
	Childcare		Kindergarten	According to an overall planning for community neighboring area
	Old age care	Meal service for the elderly	Day care center	According to an overall planning for community neighboring area
	Health service station		Community health service station	According to an overall planning for community neighboring area
	Cultural activity station		Reading room Multi-function room (chess and cards, family and children)	According to an overall planning for community neighboring area
Fitness facilities		Outdoor fitness facilities	Ring fitness footpath	In conjunction with square and green land of the community
		Playground for children		In conjunction with square and green land of the community
		Activity venue for the elderly		In conjunction with square and green land of the community
Convenience service		Public toilets、Express delivery facilities、Renewable resources recovering spot		At community entrance and exit and main passages
		Outdoor air drying facilities		In spaces of residence courtyards
Commercial outlets		Convenient store、Drug store、Laundry、Barber shop	Vegetable market、Pet store、Eatery、Bank outlet	According to an overall planning for community neighboring area
Public traffic facilities		Bus station / metro station		By streets in community neighboring area

便民设施布局应遵循“相对集中有序分散”的原则

The layout of facilities for the convenience of the people shall follow the principle of “relatively concentrated, and orderly dispersed”

- 结合小区的主要出入口、公共空间、居民楼出入口等整体考虑。
- 居民使用频率高的设施与居民出行交通流线相结合设置，使用频率一般的设施可与主要的户外活动空间结合设置，同一空间的多种设施尽量集中布局。
- 预留一定的空间，应对今后新增设施要求。

- Overall consideration is needed in conjunction with the main entrance and exit of the community, public space, and residential building entrance and exit.
- Facilities with high frequency of use can be arranged in combination with the traffic flow line of residents, facilities with general frequency of use can be arranged in main outdoor activity space, and multiple facilities in the same space should be arranged in a centralized manner as far as possible.
- Some space should be reserved for new facilities required in the future.

提升便民设施设计品质

Improve the design quality of convenient facilities

- 各类便民设施均应考虑风格与环境协调、材质耐久、工艺安全、无障碍设计，兼顾夜间照明、遮风避雨等多场景使用需求。
- 在满足实用性的基础上，适当进行艺术美化，提升小区的环境面貌。

- For all types of convenient facilities, consideration should be taken for the harmonic style with environment, material durability, process safety, barrier-free design, also taking into account night lighting, shelter from wind and rain, and other multi-scene use requirements.
- On the basis of meeting the applicability needs, they should be beautified artistically to improve the appearance of the residential area.

图2-6-1 南京金尧山庄利用闲置空地新建养老服务中心 / New elderly service center built in idle space in Jinyao Villa, Nanjing | 图片来源：编写组自摄

6.2 合理设置快递投放位置
Rationally arrange locations for express delivery

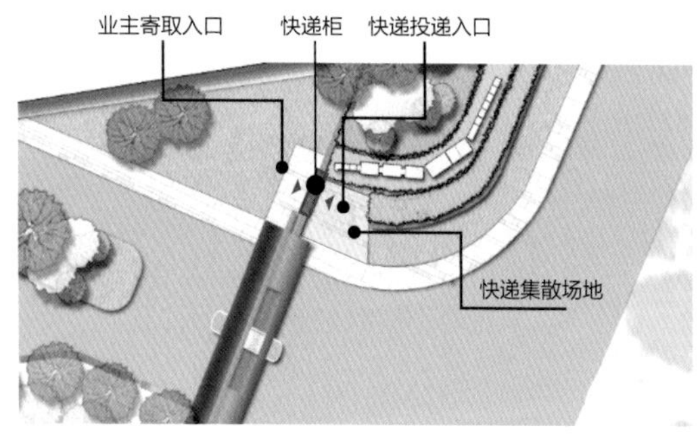

图2-6-2 小区入口“外投内取”快递柜布局 / Layout of express delivery cabinet for “external delivery and internal taking” at community entrance | 图片来源：编写组自绘

图2-6-3 小区入口“外投内取”快递柜实景图 / Actual view of express delivery cabinet for “external delivery and internal taking” at community entrance | 图片来源：枫景苑A区改造设计项目

图2-6-4 快递柜场地设计 / Express delivery ground design | 图片来源：编写组自摄

快递投放设施应满足一定服务半径
Express delivery facilities shall cover a certain service radius

· 宜结合小区出入口和主要通道布置，服务半径不超过200m。
· 有条件的小区可与门卫、围墙结合设置，采取“外投内取”的模式，方便投取。

· They should be arranged at the entrance and main passage of the community, and the service radius shall not exceed 200m.
· Where conditions are permit, they can be set up with the guard room and fence of the community, in the form of “external delivery and internal taking”, to facilitate delivery and taking.

合理设计快递投放场地
Rationally design ground for express delivery

· 快递箱柜设置的场地应硬化、排水、防滑，固定装置须牢固安全，并定期检修。
· 须考虑夜间照明设施，并应注意夜间照明及噪声对邻近居民的影响。
· 大件快递应做好防雨、防盗等措施。

· The ground for the express delivery cabinet shall be hardened, with water drainage and slip-proof, be firmly and safely installed and checked and repaired regularly.
· Night lighting facilities must be provided and attention should be paid to the impact of night lighting and noise on neighboring residents.
· Provisions against rain and theft shall be made for large mails.

6.3 设置老年助餐点
Set up meal service places for the elderly

老年助餐点配建应符合相关要求
Meal service places for the elderly shall comply with relevant requirements

- 老年助餐点应满足一定的面积要求，提供必备的冷藏、加热、分餐、就餐等设施，并符合国家规划、消防、环保等有关部门的相关要求。
- 助餐点功能布局应实用合理，无障碍设施完善，有适宜的空调等温度调节设施。有条件的可设置无障碍卫生间。
- 对于条件有限的小区，可与周边小区共享老年助餐点，但步行距离应小于300m。
- 鼓励各地探索“1+X”老年助餐服务模式，即1个老年餐集中配送和制作中心，为若干助餐点提供配送服务。

- The meal service place shall meet the area requirement, be provided with necessary facilities such as refrigeration, heating, separate meals and dining, and also meet the relevant requirements of the state planning, fire protection, environmental protection departments.
- The meal service place shall have practical and rational functions, with complete barrier-free facilities and air conditioner to regulate temperature. Barrier-free toilets shall be provided when conditions permit.
- Communities with restricted conditions may share meal service place with surrounding community, provided that the walking distance shall be less than 300m.
- It is encouraged to explore the “1+X” pattern for meal service for the elderly, i.e. one centralized making and distribution center distributes meals to a number of places.

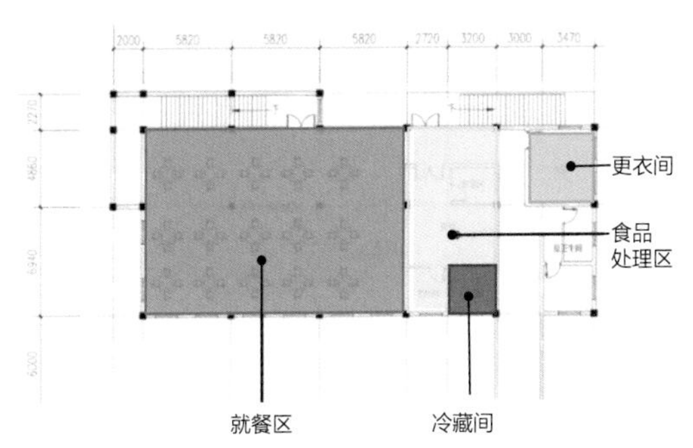

图2-6-5 老年助餐点功能布局示意图 / Function layout schematic diagram of meal service place for the elderly | 图片来源：编写组自绘

增设老年助餐点的几种方式
Ways to add meal service places for the elderly

- 结合小区公共用房改扩建增设。
- 利用建筑架空层、住宅底层改造增设。
- 利用小区空地新建，但不得影响周边建筑安全、日照等要求。

- By modifying, expanding and building public houses in the community.
- By modifying elevated layer or ground floor of residences.
- Building on idle space in the community, without affecting the safety and sunshine of buildings nearby.

图2-6-6 结合社区服务用房增设老年助餐点 / Add meal service place for the elderly in community service room | 图片来源：编写组自摄

6.4 统一布置室外晾晒设施
Arrange unified outdoor air drying facilities

合理设置晾晒场地
Rationally set up air drying ground

图2-6-7 结合住宅院落设置晾晒设施 / Air drying facilities in residence courtyard | 图片来源：编写组自摄

- 晾晒场地宜结合住宅院落、日照充足的位置布设，注意避开主要通道出入口、楼栋出入口、消防疏散通道等区域。
- 场地铺装宜采用硬质平整的铺装或耐踩踏植物，不宜采用植草砖等镂空铺装，方便老人行动。

- It should be located in residence courtyard with sufficient sunshine, and areas such as main passage entrances, building entrance and firefighting evacuation channels should be avoided.
- The ground should be paved with hard and flat materials or stamp-proof plants, hollow pavement such as grass planting bricks should not be used, to facilitate walking by the elderly.

晾晒设施形式兼顾实用与景观
The form of facilities shall be both practical and in good landscape effect

- 晾晒设施应统一设计，鼓励融入创意文化，实用与景观兼顾。
- 晾晒设施应有一定的强度和刚度，保证安全性。
- 晾晒架高度需符合晾晒衣物的需要，方便老年人使用。

- Air drying facilities should be designed in a unified way, and it is encouraged to merge creative culture, for both practicality and landscape effect.
- Air drying facilities shall have certain strength and stiffness to ensure safety.
- The height of air drying rack shall meet the needs of air drying clothes, convenient for the elderly to use.

6.5 提供方便可寻的公厕
Provide convenient public toilets

公共厕所应实现 5 分钟步行可达
Public toilets shall be accessible within 5 minutes of walking

- 公厕可结合小区入口、门卫房、公共活动空间、社区用房等设置，方便使用。
- 独立设置时应合理选址，并做好邻近住户的沟通工作，邻避处置，造型风格应与小区环境协调。
- Public toilets can be set up in combination with the community entrance, gatehouse, public activity space and community room for convenient use.
- For independent toilets, the site shall be selected rationally, and proper communication shall be made with neighboring households for nimby setup, and the style should be harmonic with the community environment.

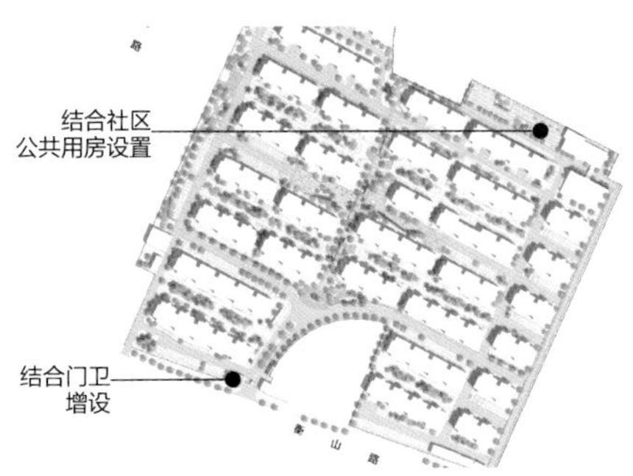

图2-6-8 昆山中华园北村公厕布局 / Public toilet layout in Zhonghua Garden North, Kunshan | 图片来源：昆山中华园北村老旧小区改造项目成果

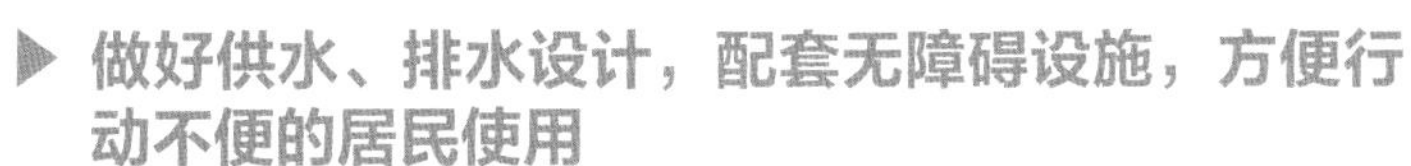

做好供水、排水设计，配套无障碍设施，方便行动不便的居民使用
Make proper water supply and drainage design, provide barrier-free facilities to be convenient for residents with limited mobility

- 完善无障碍设施，墙上应安装不同高度的安全扶手。
- 安全扶手的颜色应与墙体颜色形成强烈对比，照顾视力减退的人。
- Complete barrier-free facilities, safety handrails shall be fixed on wall at different heights.
- The color of the safety handrail should be in sharp contrast with the wall color, for use by people with impaired vision.

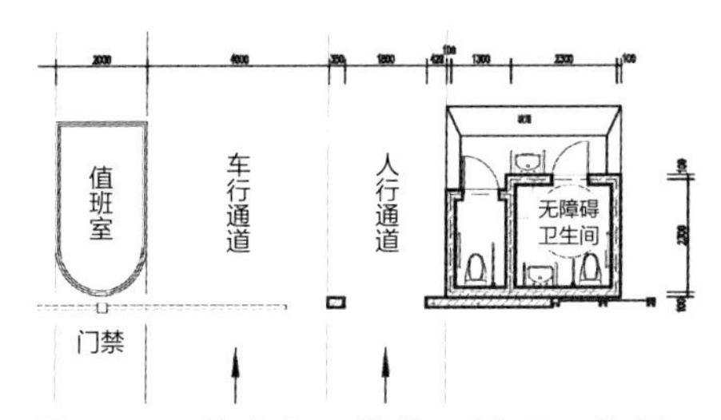

图2-6-9 结合入口增设无障碍卫生间 / Barrierfree toilets nearby entrance | 图片来源：昆山中华园北村老旧小区改造项目成果

图2-6-10 小区公厕设计应与周边环境协调 / Design of public toilets shall be harmonic with surrounding environment | 图片来源：中华园东村老旧小区改造项目成果

6.6 为居民和来访者提供清晰的指引标识
Provide clear guide signs for residents and visitors

合理设置方向指引标识
Rationally set up direction guide signs

- 小区入口应结合进入方向设置小区平面图。
- 主要道路转弯处应设置方向指示牌。
- 停车场应设置出入口标识、无障碍车位标识。
- A plot plan of the community shall be set up in the entering direction of the entrance.
- Direction signboard shall be provided at turns of main road.
- Signs of entrance and exit and for barrier-free position shall be provided in the parking lot.

楼栋标识应易于辨识
Building signs shall be easily identifiable

- 楼栋编号宜设置于山墙，字体、高度等要求易于辨识。
- 楼栋单元入口编号应设置在入口醒目处，便于找寻。
- Building number should be marked on the gable, in font and height easy to identify.
- The entrance number of the building unit shall be set in an eye-catching place at the entrance for easy searching.

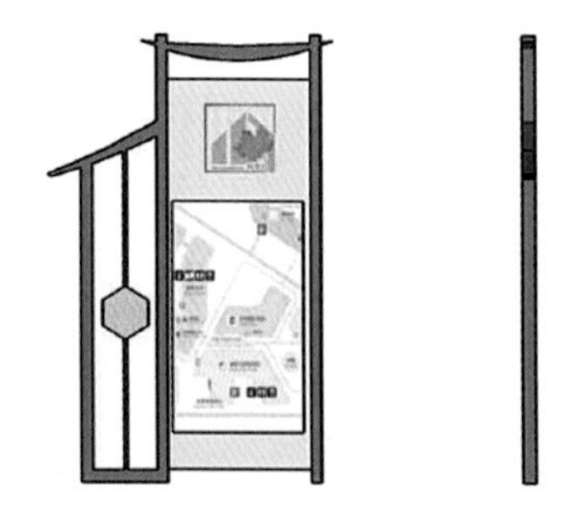

图2-6-11 小区总览图 / Community overview
| 图片来源：枫景苑A区改造设计项目

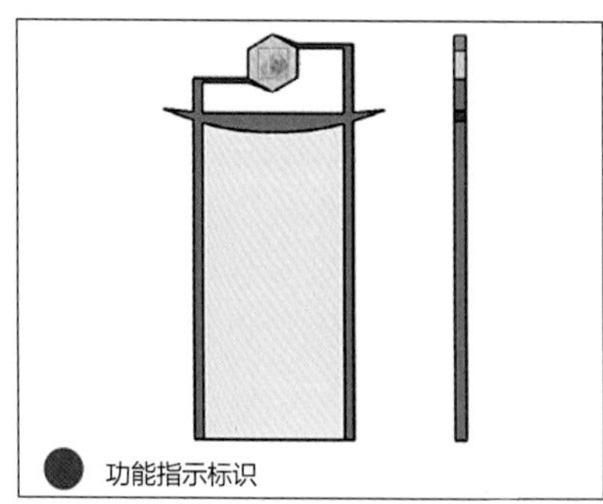

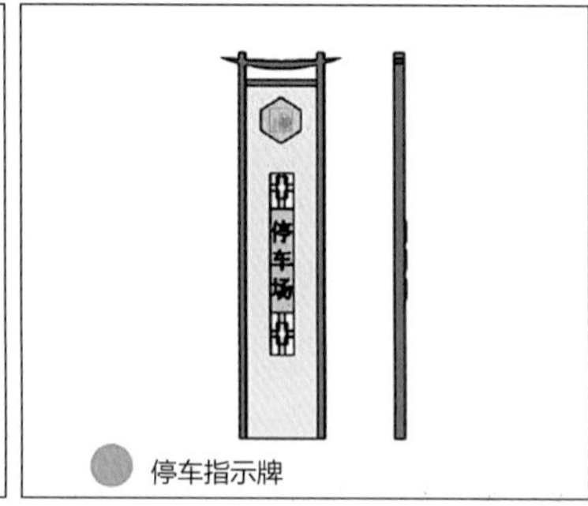

图2-6-12 停车场出入口标识 / Entrance and exit signs of parking lot | 图片来源：枫景苑A区改造设计项目

公共场所应完善标识规范行为
Set up complete signboards in public venues for proper behavior

· 公共场所根据需要设置安全警示牌、温馨提示牌、无障碍设施标识牌等，提醒活动人群注意。

· In public venues, safety warning signs, warm prompt signs and signs for barrier-free facilities should be set up to remind the people for attention.

标识设计注意事项
Precautions in sign design

· 各类标识牌布局应做到位置醒目，且不对行人交通和景观环境造成妨碍。
· 标识牌设计应简洁、美观、表达清晰，尺寸大小应符合辨识、审美要求，风格统一具有系列性，且与小区整体建筑环境风貌相和谐。
· 标识应有足够照明或内置光源，满足夜间使用。

· All signs should be set at eye-catching locations, and not interfere with pedestrian traffic and landscape environment.
· Signboard design shall be simple, beautiful, and clear in expression, in sizes conforming to identification and aesthetic requirements, in unified style as a series, and harmonic with the overall architectural environment of the community.
· Signs should have sufficient lighting or built-in light source, to be visible at night.

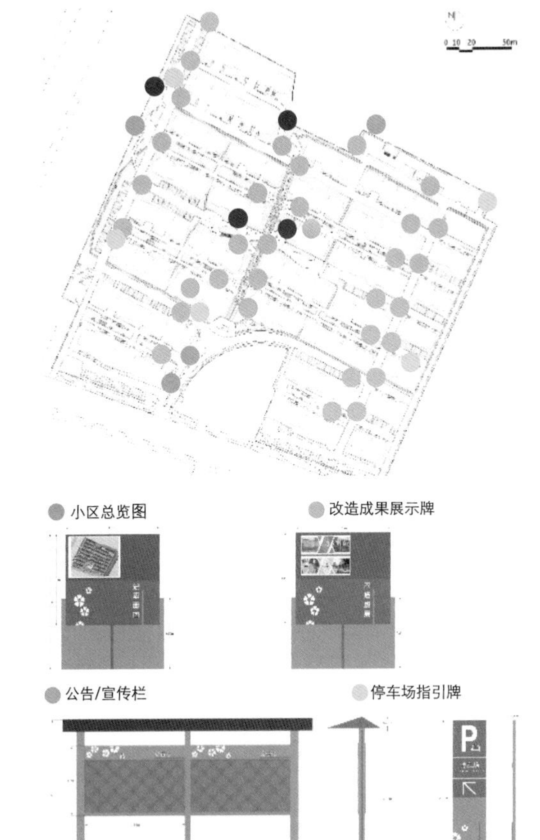

图2-6-13 昆山中华园北村标识系统设计 / Design of sign system in Zhonghua Garden North, Kunshan | 图片来源：昆山中华园北村老旧小区改造项目成果

· 标识系统进行系列化设计。
· 设计立意呼应小区整体风貌和色彩环境。
· 成为小区环境的有机组成部分。

· The sign system shall be designed as a series.
· The design echoes the overall style and color environment of the residential area.
· It shall become an organic part of the community environment.

6.7 有条件的小区可设置老年日间照料中心
Set up day care centers for the elderly in communities with ready conditions

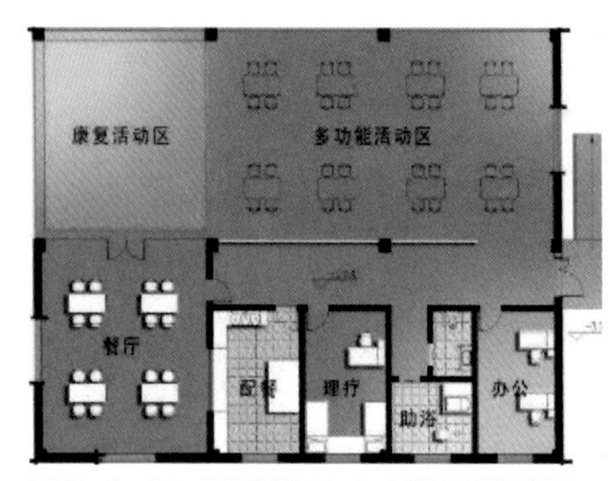

图2-6-14　日间照料中心功能布局示意图 / Function layout schematic diagram of a day care center | 图片来源：编写组自绘

▶ 配置功能完善的日间照料中心
Day care center with complete functions shall be provided

· 日间照料中心主要包括休息室、沐浴间和厨房、餐厅（没有条件时可与周边餐饮单位合作，小区内仅设餐厅）、康复保健用房、活动中心以及办公室、公共卫生间、其他用房等，可结合实际需求配置。

· A day care center mainly includes lounge, bath room, kitchen, dining hall (if conditions are not available, cooperate with catering units nearby and only set up a dining room in the community), rehabilitation and health care room, activity center, office, public toilets and other rooms, etc., and they can be set up according to actual needs.

▶ 日间照料中心宜选择日照充足、通风良好、交通便利的区域设置
Daytime care centers should be located in areas with sufficient sunshine, good ventilation and convenient transport

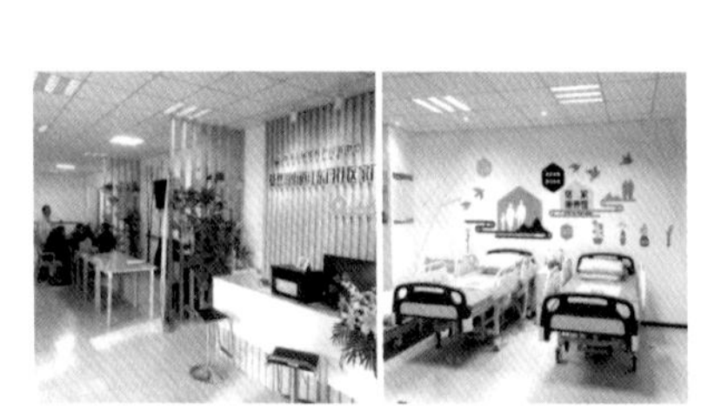

图2-6-15　南京江东门社区日间照料中心 / The day care center in Jiangdongmen Community, Nanjing | 图片来源：编写组自摄

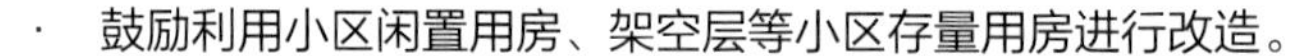

· 鼓励利用小区闲置用房、架空层等小区存量用房进行改造。

· 满足日照、消防等卫生安全要求，多层建筑应配有电梯。

· 厨房配餐应满足相关规范要求，不得影响周边居民生活。

· It is encouraged to renovate existing rooms such as unused rooms and elevated layer of floor for this purpose.

· Sunshine, fire protection safety requirements shall be met, and multi-storey buildings shall be equipped with elevator.

· Kitchen catering shall meet the relevant standard requirements, and must not affect the life of residents nearby.

6.8 适当预留场地布设临时便民设施
Make appropriate reservation for temporary facilities to facilitate people

▶ 结合小区出入口、活动广场预留临时便民场地
Reserve temporary convenient venues for people at community entrances and exits and squares

· 结合小区出入口、公共活动广场预留场地，方便布设临时便民设施，或开展便民活动，如修补衣物、修鞋、磨刀、理发、义诊、二手集市等项目。

· Areas shall be reserved at community entrances and exits and public squares, to set up temporary convenient facilities or convenient activities, such as clothes patching, shoe repair, knife sharpening, hair-cutting, voluntary clinics and second-hand market.

▶ 引导便民场地规范发展
Guide the formalized development of convenient venues

· 场地设计应兼顾日常使用与临时特定功能使用要求，适应需求变化。

· 物业应结合便民活动做好场地管理，保障临时设施有序布设、环境整洁。

· The venue design should take into account the daily use and temporary use for specific purposes, to adapt to the changing needs.

· The property management shall well manage the venues in combination with the convenience activities to ensure orderly arrangement of temporary facilities and clean environment.

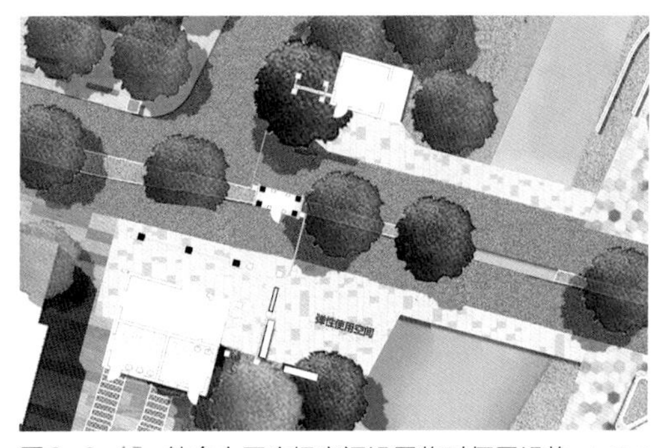

图2-6-16 结合小区广场空间设置临时便民设施 / Set up temporary convenience facilities in space of community square | 图片来源：枫景苑A区改造设计项目

6.9 挖掘周边资源，进一步完善社区服务
Tap surrounding resources to further consummate community services

挖掘梳理闲置畸零地
Tap and sort out idle corner land

- 高架桥下用地
- 废弃仓储、工厂用地
- 后退道路红线用地
- 边角畸零地
- 公共建筑屋顶空间

- Land under a viaduct
- Land of discarded warehouse and factory
- Land in property line area of retreated road
- Edge and corner land
- Space on roof of public building

· 候学区的改造，让家长们不再烈日暴晒，风吹雨淋，宜人的休憩设施，适老化的设计、儿童趣味空间，处处体现出人性的关怀。

· Livable school district, parents will not be exposed to the hot sun, wind and rain, pleasant rest facilities, suitable for aging design, children's fun space, everywhere reflects the care of human nature.

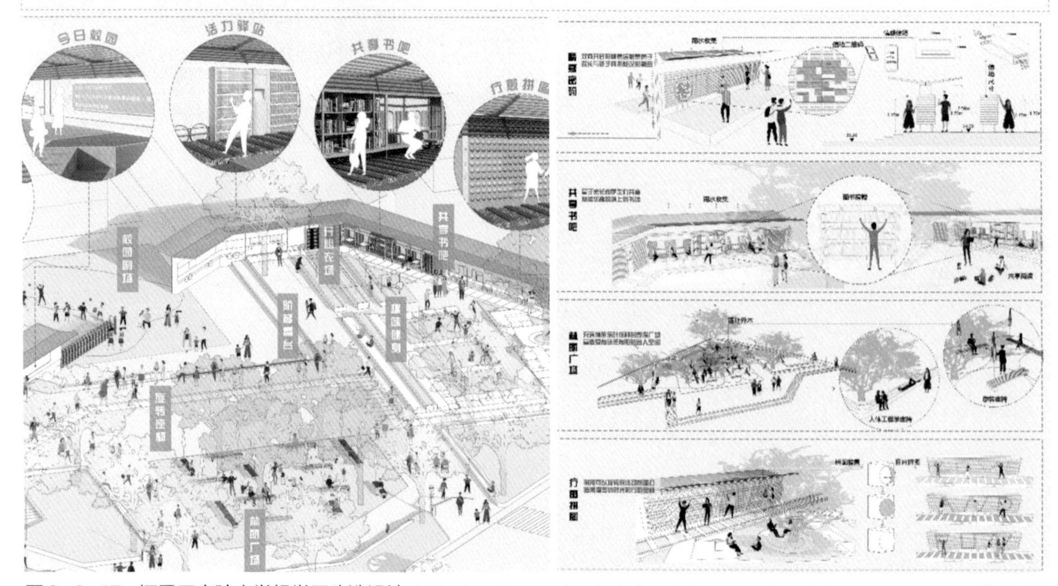

图2-6-17　栖霞区实验小学候学区改造设计 / Design of hou school district of qixia experimental primary school | 图片来源：编写组自绘

2000m^2 以下的闲置用地利用方向
Utilization orientation of idle land less than 2000m^2

· 规模较小的闲置地块可用于完善停车设施、街角游园、健身场地（器材）、小型便民或市政设施。

· Idle land of small size can be used to improve parking facilities, street corner parks, fitness venues (equipment), small amenities or municipal facilities.

· 外立面采用与住宅立面相呼应的红砖。
· 采用玻璃砖镶嵌图案，同时解决采光问题。
· 砖墙还可作为住区的宣传橱窗。
· The exterior is made of red bricks that echo the facade of the house.
· Glass brick mosaic pattern is adopted, also solving the problem of lighting.
· The brick walls can also serve as a publicity window for the community.

图2-6-18 边角地改造成自行车棚 / The corner land has been converted into a bicycle shed | 图片来源：编写组自摄

· 带状空间分成通行区和休闲区两部分。
· 休闲区设置有遮阴廊架、儿童活动场地、健身器材和休憩桌椅，提高空间使用的复合化。
· The strip space is divided into two parts: passage area and leisure area.
· The leisure area is equipped with a shade gallery frame, children's activity space, fitness equipment and tables and chairs for rest to enhance the compound use of space.

图2-6-19 小区之间的带状闲置地改造为街角游园 / A strip of vacant land between communities has been transformed into a street corner garden | 图片来源：编写组自摄

图2-6-20 小区周边的闲置地改造为健身活动场地与社区农园 / The idle land around the community has been transformed into fitness activity grounds and community plantation garden | 图片来源：编写组自摄

6.9 挖掘周边资源，进一步完善社区服务
Tap surrounding resources to further consummate community services

2000m² 以上的闲置用地利用方向
Utilization orientation of idle land over 2000m²

- 社区用房
- 社区卫生服务站
- 便利店
- 幼儿园
- 社区公园
- 健身场地（球类等）

- Community management room
- Community health service station
- Convenient store
- Kindergarten
- Community park
- Fitness ground (balls and so on)

共享体育场地

· 原小区内篮球场使用率不高，开放后成为街区内居民共享设施。

Shared sports ground

· The utilization rate of the original basketball court in the community was not high. After opening, it became a shared facility for residents in the block.

改造街角公园

· 利用建筑后退道路红线空间改造街角公园。

· 公园内植入儿童活动乐园，与周边小学共享。

Renovation of street corner park

· A street corner park has been built in the property line space of building retreat from road.

· A children's activity park is implanted in the park and shared with the nearby primary school.

新建社区服务中心

· 利用闲置的底层商业空间改造为社区服务中心。

· 提供阅览、活动等功能场所，满足居民文化休闲需求。

Newly built community service center

· The idle commercial space on the ground floor has been transformed into a community service center.

· Provide reading, activities and other functional venues to meet the cultural leisure needs of residents.

新增菜市场

· 利用闲置建筑改造菜市场。

· 植入便民服务设施，形成功能复合的公共场所。

New vegetable market

· An idle house has been transformed into a vegetable market.

· The convenience service facilities are implanted to form a public place with multiple functions.

图2-6-21 南京姚坊门宜居街区闲置空间利用 / Utilization of idle space in Yaofangmen Community, Nanjing | 图片来源：编写组自摄、自绘

6.10 提供服务便利的商业网点
Provide convenient store and service spots

5~10 分钟步行可达的便民商业设施
Convenient commercial facilities accessible with 5~10 minutes walk

· 菜市场，银行、电信等营业网点，超市/便利店，药店，洗衣店，理发店，餐饮店，宠物店。

· Vegetable market，Bank and telecom outlets，Supermarket/convenient store，Drug store，Laundry，Barber shop，Eatery，Pet store.

鼓励便民商业沿街线型布局，形成活力街道空间
Encourage convenient commerce to arrange along the street, to form a vitality street space

· 结合社区生活性街道，利用沿街建筑底层布局，方便可达。

· 沿街线型布局，有利于提升街道活力，让居民享受社区生活。

· Use the ground floor of buildings along street in the living street in the community, convenient and accessible.

· Arrangement along the street can enhance the vitality of street and let residents enjoy community life.

发挥市场作用，推动社区服务内容和形式的多样化、供给的灵活化以及服务的精准化，适应生活需求和行为方式的变化
Give play to the role of the market, promote the diversification of community service content and forms, flexible supply and accurate services, to adapt to the changes in life demands and behavior

· 苏宁小店围绕用户和用户家庭的“厨房”，致力打造成为每个社区的“共享冰箱”，并为消费者提供其他增值服务。

· Suning stores focus on creating “kitchen” of users and their families, strive to become the “shared refrigerator” of every community, and also provide other value-added services for consumers.

· 南京携才养老连锁机构提供社区居家养老、文化娱乐、健康管理、康复理疗等服务。

· Xiecai Old-age Care chain institutions in Nanjing provide community home care, cultural entertainment, health management, rehabilitation therapy and other services.

图2-6-22 小区入口边的银行、便利店 / Bank and convenient store by the community entrance | 图片来源：编写组自摄

图2-6-23 小区周边功能复合的商业街 / Shopping street with compound functions by the community | 图片来源：编写组自摄

图2-6-24 阿里菜鸟驿站提供便民社区服务 / Alicainiao post station provides convenient community services | 图片来源：编写组自摄

6.11 提供就近的社区卫生服务
Provide community health service nearby

满足 10~15 分钟步行可达的服务半径
Service radius accessible with 10~15 minutes walk

· 可以结合社区中心布局，也可利用存量空间独立布局，应方便可达。

· It can be arranged either in the community center, or independently set up in the existing space, and should be convenient and accessible.

图2-6-25 上海邻里汇 / Shanghai community care | 图片来源：编写组自摄

设置生活服务与医疗健康服务，包括健康管理驿站、聊天吧、老年助餐点、社区卫生服务站等功能。
Provide life service and medical and health service, including health management station, chat bar, meal service for the elderly, community health service station and other functions.

图2-6-26 上海怡乐家园照护站 / Shanghai Yilejiayuan nursing station | 图片来源：编写组自摄

设置提供养老服务与休闲活动的照护站，老年人可享受长者照护、日托照料、老年助浴、康复健身等服务。
A care station is set up to provide old age care service and leisure activities, and the elderly can enjoy such services as elderly care, day care, bath assistance, rehabilitation and fitness.

提供亲民的医疗保健服务
Provide readily accessible health care services

· 满足门诊医疗、康复保健、便民取药、居民体检等服务需求。

· Meet the needs of out-patient medical treatment, rehabilitation and health care, convenient access to medicine, physical examination for residents and other services.

图2-6-27 南京宁夏路街区内便民医疗点 / Neigh borhood medical facility at Ningxia road block in Nanjing | 图片来源：编写组自摄

图2-6-28 居民小区附近可就诊 / Medical service available near residential areas | 图片来源：编写组自摄

6.12 提供 5 分钟可达的公交站点
Arrange bus stops accessible in 5 minutes

公交站点与住区出入口合理衔接
Rational connection of bus stops with community entrance and exit

· 公交站点靠近住区出入口，方便居民出行。

· 公交站点与步道系统紧密连接，方便绿色出行。

· Bus stations should be close to community entrance and exit, to facilitate residents.

· Bus stops shall be closely linked with the footpath system, to facilitate green travel.

智慧公交站点提供多元便利服务
Smart bus station providing diversified convenient services

· 提供可实时查询公交动态信息的站牌。

· 设计具有本地特色的公交站台。

· 增加舒适性的休息座椅。

· 增加社区信息或文化展览平台。

· Station screen providing real-time bus movement information.

· Design bus stations with local features.

· Provide rest seat for more comfort.

· Add community information or culture exhibition platform.

雨篷为乘客遮阳避雨

墨水屏实时显示车辆到达信息

电子广告牌滚动播放新闻与公益广告

各条公交线路信息

供乘客等候的座椅

高清监控、垃圾箱、手机快充等其他功能

图2-6-29 智能公交站 / Smart bus station
| 图片来源：编写组自摄、自绘

Construction Guidance for
Renovation of Old Communities in
Jiangsu

江苏老旧小区改造建设导引

7

以人为本改善公共活动空间
Improve Public Activity Space with People as the Core

7.1 配建一定面积的室外活动场地
Provide outdoor activity venues with fair space

充分挖掘零散用地增设公共活动空间
Make full use of scattered land for more space for public activities

· 利用院落背向空间、宅旁、边角畸零地改造为公共活动空间。
· 改造后的活动广场、绿地应与步道系统衔接。

· Space on the back of courtyard, by houses and edge and corner land can be turned into space for public activities.
· The activity square and green space after renovation should be connected with the footpath system.

小区主要的公共空间应具有一定规模
Main public space in a community shall have a fair size

· 主要公共活动空间面积不宜小于600m^2。
· 合理划分功能分区，形成功能复合的活动场所。
· 保证一定的硬质铺装区，便于居民开展活动。

· The area of main public activity space should not be less than 600m^2.
· Divide function zones reasonably, to form activity venues with compound functions.
· Ensure some hard pavement area, to facilitate residents in activities.

结合公共空间设置健身场地
Set up fitness venues in public space

· 健身场地周边应合理布局绿化，降低噪声对居民的影响。
· 场地铺装应选用弹性缓冲材料。

· There shall be rationally planted around fitness venues, to reduce the impact of noise on residents.
· The pavement shall be of elastic and buffer materials.

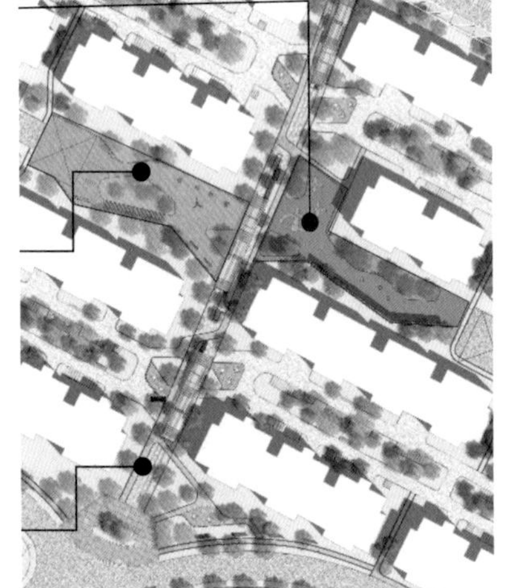

图2-7-1 利用宅旁等零散用地改造为公共活动空间 / Renovate scattered land by houses into public activity space | 图片来源：昆山中华园北村老旧小区改造项目成果

公共空间应提供座椅等人性化设施

Public space shall be provided with humanized facilities such as seats

- 结合广场活动功能，提供休息座椅、遮阳避雨、公厕、垃圾分类回收等人性化设施，合理布局。
- 对公共活动空间周边的干扰因素，如停车位、垃圾箱等位置进行调整优化。

- In combination with the activity function of square, humanized facilities such as rest seats, sunshade and rain shelter, public toilets and garbage classification and recycling are provided with rational layout.
- Adjust and optimize the disturbance factors around the public activity space, such as parking positions and garbage bins.

图2-7-2 公共空间提供休闲座椅与遮阴设施 / Public space shall be provided rest seats and shade facilities | 图片来源：编写组自摄

南京天顺苑小区宅间健身场地改造：
保留原有乔木；增设健身设施；辅以环境小品和休息设施。

Renovation of fitness venue between houses in Tianshun Garden, Nanjing:
Keep the original trees; add fitness facilities; supplemented with environmental vignettes and rest facilities.

图2-7-3 南京天顺苑小区宅间健身场地改造 / Renovation of fitness space between houses in Nanjing tianshunyuan community | 图片来源：编写组自摄

7.2 提供老人、儿童活动场地
Provide activity venues for the elderly and children

▶ 儿童活动场地应满足安全性、趣味性要求
The playground for children should meet the requirements of safety and being interesting

- 儿童活动区宜结合小区绿地、广场等公共空间设置，具有良好的日照、通风条件，远离交通干扰；同时应与住宅楼保持一定距离，避免活动声音对周边居民的干扰。
- 活动场地铺装与设施选材应确保环保和牢固，场地应保持视线通透。应考虑一定的家长等候看护区域，设置座椅、遮阴花架等设施。
- 儿童活动区设计要有趣味性，挖掘场地特点，设计符合儿童尺度和审美的空间环境。
- Children's activity areas should be set in the community green space, square and other public spaces, with good sunshine and ventilation conditions, away from traffic interference; also, it should keep a certain distance from residential buildings to avoid the interference of the activity sound to the surrounding residents.
- The paving and materials selected for the activity venues shall ensure environmental protection and firmness, and good line of sight shall be maintained on ground. Some waiting and caring areas for parents shall be considered, provided with facilities such as seats and shade flower racks.
- Children activity area shall be designed interesting, with its features and space environment suiting the dimensions and aesthetics of children.

图2-7-4 儿童活动场地设计 / Design of playground for children | 图片来源：姚坊门省级宜居示范街区规划和设计服务项目

昆山中华园北村儿童活动场地改造：
利用宅旁消极空间，改造为公共活动场地。
置入儿童活动休闲设施。
提供休息座椅，打造趣味景墙。
Renovation of playground for children in Zhonghua Garden North, Kunshan:
Renovate idle space by houses into public activity venue.
Activity and leisure facilities for children are included.
Rest seats and interesting landscape wall.

图2-7-5 儿童活动场地设计 / Design of playground for children | 图片来源：编写组自摄

图2-7-6 昆山中华园北村儿童活动场地改造 / Reconstruction of children's playground in North Village of Zhonghua garden in Kunshan | 图片来源：昆山中华北村小区改造规划项目

老人活动场地应提供遮阳避雨设施

Sunshade and rain shade facilities shall be provided for activity venues for the elderly

- 宜结合小区绿地、广场等公共空间设置，具有良好的日照、通风条件。
- 结合老人兴趣，提供健身、棋牌等活动场地，合理分区。
- 提供休息以及遮阳避雨设施。
- 场地高差处理应采用坡道，完善无障碍设施。

- Should be set in the community green space, square and other public spaces, with good sunshine and ventilation conditions.
- Provide fitness and chess and card activity venues with rational zonation according to interests of the elderly.
- Rest and sunshade and rain shade facilities shall be provided.
- Ramps shall be provided for ground at different heights, and barrier-free facilities shall be completed.

图2-7-7 公共空间改造 / Public space transformation | 图片来源：编写组自摄

南京天顺苑小区老年休闲场地改造：
利用宅间空间改造为老人休闲交流场所，方便老人下楼活动，设置木质休息座椅及遮阴花架，利用低矮灌木绿化围合营造场所感。

Renovation of leisure venue for the elderly in Tianshun Garden, Nanjing:
Space between houses has been renovated into leisure and communication venue for the elderly, to facilitate activities of the elderly outdoor, wooden rest seats and shade flower racks are provided, and low shrubs are plants around to make it a proper venue.

图2-7-8 老人活动场地设计 / Design of activity venue for the elderly | 图片来源：姚坊门省级宜居示范街区规划和设计服务项目

7.3 完善无障碍设施 Complete barrier-free facilities

图2-7-9 塑胶铺装 / Plastic pavement | 图片来源：编写组自摄

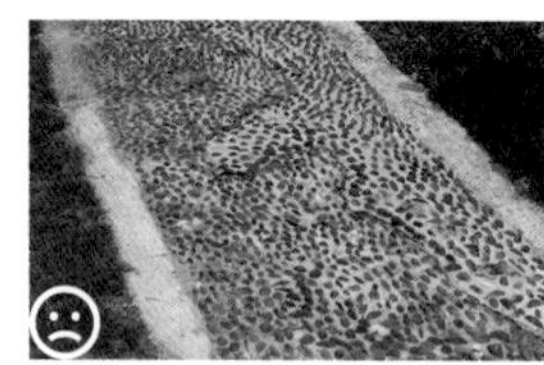
图2-7-10 卵石铺装 / Pebble pavement | 图片来源：编写组自摄

图2-7-11 透水砖铺装 / Permeable brick pavement | 图片来源：编写组自摄

图2-7-12 汀步石铺装 / Step stone pavement | 图片来源：编写组自摄

昆山中华园北村单元入口无障碍改造：
入口处坡道优化，降低坡道斜度。
入口一侧位置布置休闲座椅，方便老人及残疾人临时停靠。
Barrier-free renovation for unit entrance in Zhonghua Garden North, Kunshan:
Optimize the entrance ramp by reducing the sloping.
Seat arranged at a side of the entrance to facilitate temporary stop by the elderly and disabled.

图2-7-13 昆山中华园北村单元入口无障碍改造 / Barrier free renovation of North Village unit entrance of Zhonghua garden in Kunshan | 图片来源：昆山中华北村小区改造规划项目

▶ 小区应形成完整的无障碍步道系统
A complete barrier-free footpath system shall be formed in the community

- 无障碍步道应连接小区各类活动场地、公共设施入口、住宅单元入口、停车场等，并与城市道路步行系统无障碍衔接。
- 小区主路人行道应设置盲道，盲道位置尽量避免与下面管线重合，减少检查井对盲道的影响。
- 小区入口、道路交叉口均应设置缘石坡道，保障轮椅通行。
- 小区步道应满足无障碍要求，尽量避免卵石、汀步等形式，并采用防滑材质。

- Barrier-free footpaths shall be connected with all kinds of activity venues, entrances of public facility, residential units and parking lots, etc., and be barrier-free with the urban road walking system.
- Path for the blind should be set up on the main sidewalks of the community, and the location of the path shall not coincide with the pipelines below, so as to reduce the influence of inspection wells on the path.
- Curb ramps shall be set up at the entrance of the community and at the intersection of roads to ensure wheelchair access.
- The footpath in the community shall meet the barrier-free requirements, avoid pebbles and steps as far as possible, and be built with anti-slip material.

▶ 建筑出入口均应设置坡道
Ramps shall be provided at building accesses

- 结合周边空间可以采取多种形式，如一字形、U形等。
- 坡道的坡度、宽度以及建筑入户门的宽度应符合相关规范标准，方便轮椅通过。

- A number of forms can be adopted, such as straight and U-shaped according to the surrounding space.
- The slope and width of the ramps, and the width of the building entrance doors, shall comply with relevant codes and standards for wheelchair access.

适当设置无障碍停车位
Set up barrier-free parking positions as appropriate

· 根据规范要求，设置相应比例的残障人士停车位，位置应方便出入。

· According to the specification, parking for the disabled shall be arranged in certain proportion at convenient locations for access.

完善无障碍设施的标识
Complete signs for barrier-free facilities

· 结合无障碍设施，于显著位置设置无障碍标识，保障残障人士使用需求。

· For barrier-free facilities, barrier-free signs shall be set up at conspicuous locations to ensure use by the disabled.

公共走廊、坡道应设置双层扶手
Double level handrails shall be set in public corridors and ramps

· 考虑残疾人使用要求，走廊、坡道应设置双层扶手，扶手高度满足轮椅使用要求。

· Considering the use requirements of the disabled, the corridors and ramps should be provided with handrails at double levels, and the height of the handrails should meet the use requirements of wheelchairs.

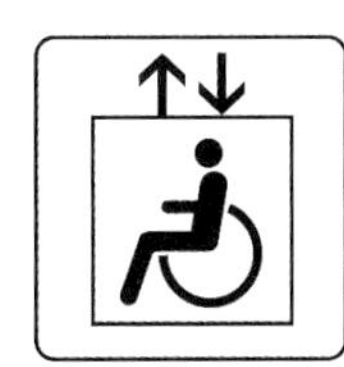

图2-7-14 无障碍标识 / Barrier-free signs | 图片来源：编写组自绘

7.4 公共空间硬件设施应符合人性化要求
Hardware facilities in public space shall comply with humanized requirements

座椅：提供温暖的座面和靠背
Seats: with warm seat surface and back

- 小区人行道、散步道沿线应布置休息座椅，间距不宜大于100m。步道宽度大于1.5m时，休息设施可沿步道一侧布置；步道宽度小于1.5m时，休息设施应结合沿线空间呈凹入式设置，便于步道通行。
- 户外座椅应充分考虑老人使用要求，尽可能提供有靠背及扶手的座椅，座椅高度不宜过高（建议35 ~ 40cm），宜选用木质或其他温暖的材料，尽量避免使用石材、金属等缺乏温度的材料。
- 休息座椅宜和高大乔木组合设置，或结合休息设施提供遮阳廊架。
- 鼓励创意设计，结合花池、树木等要素，增加设计趣味性，吸引人群休憩交往。

- Rest seats should be arranged along the sidewalks and promenade, and the spacing should not exceed 100m. When the width of the path is more than 1.5m, the rest facilities can be arranged along one side. When the width is less than 1.5m, the rest facilities should be set in a recessed form as combined with the space along the path, to facilitate the passage of the path.
- Outdoor seats should be provided with backrest and armrests as far as possible, and the seat height should not be too high (35~40cm is recommended), taking into account the use by the elderly. Wood or other warm materials should be selected, and materials lacking temperature, such as stone and metal, should be avoided as far as possible.
- Rest seats should be arranged in combination with tall trees, or sunshade corridor frame be provided for rest facilities.
- Creative design is encouraged, to combine with flower pond, trees and other elements, increase the design interest, and attract people to have a rest and for communication.

图2-7-15 座椅沿人行道一侧布置 / Seats arranged along one side of sidewalk | 图片来源：编写组自绘

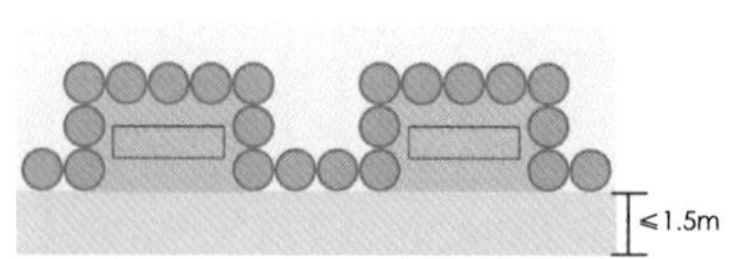

图2-7-16 座椅沿人行道凹入式布置 / Seats recessed along one side of sidewalk | 图片来源：编写组自绘

图2-7-17 创意座椅 / Seat with creative design | 图片来源：编写组自摄

图2-7-18 木质温馨座椅 / Cozy wooden seat | 图片来源：编写组自摄

花坛：避免尖锐的转角和边缘
Flower beds: avoid sharp corners and edges

- 花坛是小区公共空间常见的景观设施，需要考虑儿童活动特点，造型应避免设计成锐角，建议采用倒圆角设计。
- 在满足功能的前提下，兼顾景观品质和艺术气息，营造温馨的人文环境。
- Flowerbeds are common landscape facility in the public space of a community. It needs to consider the characteristics of children activities, and the design should avoid sharp corner. It is suggested to adopt the chamfer corner design.
- Consideration shall be given to the quality of landscape and artistic atmosphere while satisfying functional requirements, to create a warm humanistic environment.

图2-7-19 废旧轮胎制作的花坛 / Flower bed made of waste tyre | 图片来源：编写组自摄

高差处理：衔接宜舒缓和防滑
Treatment for different heights: connections should be gentle and anti-slippery

- 公共空间10~15cm的高差应至少设置2级踏步；低于10cm的高差宜采用坡道处理；台阶处可用标识或警示条提示，台阶材质应考虑防滑面。
- 小区主要活动场地需有方便轮椅及婴儿车进出的无障碍坡道。
- In a public space, at least 2 steps shall be set for height difference of 10~15cm; and a ramp should be made for height difference of less than 10cm; steps can be prompted with signs or warning strips, and steps shall have non-slip surface.
- In main activity venues in a community, barrier-free ramp facilitating access of wheelchairs and prams shall be arranged.

图2-7-20 户外台阶应舒缓设置 / Outdoor steps shall be gentle | 图片来源：编写组自摄

7.4 公共空间硬件设施应符合人性化要求
Hardware facilities in public space shall comply with humanized requirements

选择安全经济、易维护的景观材料
Select safe and economic landscape materials easy to maintain

- 安全舒适的景观材料：小区所用的景观材料应为老人和儿童营造安全舒适的户外活动空间，如彩色塑胶、彩色防滑路面、塑木等。
- 生态环保的景观材料：实现低成本景观营造，减少石材等不可再生材料的应用，可用水泥制品代替，如PC仿石砖、水洗石、水磨石等；采用透水材料，如透水混凝土砖、透水PC仿石砖、透水彩色混凝土、透水彩色沥青等，符合海绵城市建设理念。
- 避免使用的材料：为保障老人、儿童户外活动的安全性，场地铺装应防滑，避免使用抛光石材、光滑水磨石、易长青苔的青砖、瓷砖等材料。玻璃、散置碎石、散置卵石等有安全隐患的材料，不宜用于儿童活动区域。
- Safe and comfortable landscape materials: the landscape materials used in the residential area shall create safe and comfortable outdoor activity space for the elderly and children, such as color plastics, color non-slip pavement, plastic wood, etc.
- Ecological and environment-friendly landscape materials: to achieve low-cost landscape construction, reduce the use of non-renewable materials such as stone, cement products can be used instead, such as PC imitation stone bricks, washed stone, terrazzo, etc.; the use of pervious materials, such as pervious concrete bricks, pervious PC imitated stone bricks, pervious color concrete, pervious color asphalt and so on, conforms to the concept of sponge city construction.
- Materials to be avoided: to ensure the safety of the elderly and children in outdoor, the pavement should be non-slippery, and materials such as polished stone, smooth terrazzo, black bricks easy to grow moss, and ceramic tile should be avoided. Glass, scattered macadam or pebbles and other materials with potential safety hazards should not be used in children's activity areas.

彩色塑胶：安全舒适有弹性，施工便捷，适用于健身步道和儿童游戏区。
Color plastics and rubber: safe, comfortable, flexible, and convenient for construction, suitable for fitness footpath and children's play areas.

塑木：一种再生的仿木材料，兼具塑料的耐水防腐和木材的质感特性。
Plastic wood: a regenerated wood imitation material with both the water-resistant and anti-corrosion properties of plastic and the textural properties of wood.

图2-7-21 安全舒适的景观材料 / Safe and comfortable landscape materials | 图片来源：编写组自摄

7.5 利用畸零地打造小微活动场所
Make use of odd areas to build small and micro activity venues

挖掘小区低效闲置空间
Make use of idle space in the community

- 梳理小区内低效闲置空间：包括无绿化、无设施的边角零畸地块；设施陈旧简陋、难以使用的广场绿地；闲置的底层架空空间、公共建筑及其屋顶等。
- 结合低效空间位置、规模、形态等条件，合理利用，增加居民户外休闲活动场所。

- Sort out idle spaces in the community: including the edge and margin land with no plantation and facilities; square green land with old and primitive facilities and difficult to use; idle elevated space of the ground floor, public building and its roof.
- Make rational use of them according to their location, size and form, to get more outdoor leisure and activity venues for residents.

低效空间改造应征得周边居民同意
Consent shall be obtained from surrounding residents for renovation of idle space

- 充分了解周边居民真实需求，制定改造利用方案。
- 改造方案应征求周边居民、利益相关者意见，达成共识。
- 改造方案不应对周边居民日常生活、通风、采光等带来负面影响。

- Real demand of surrounding residents should be fully known, to work out renovation and utilization scheme.
- For the renovation scheme, opinions should be solicited from surrounding residents and stakeholders to reach consensus.
- The renovation scheme should not bring negative effects on the daily life, ventilation and lighting of surrounding residents.

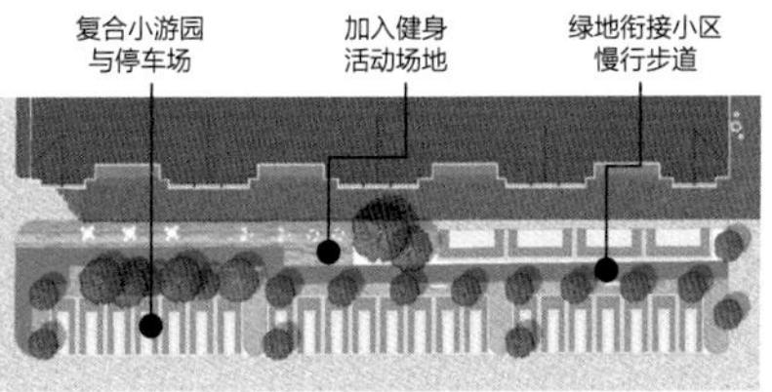

图2-7-22 老旧小游园集约化改造设计 / Intensified renovation design for old small gardens | 图片来源：编写组自摄、自绘

7.6 建设小区环形健身步道
Build ring fitness walkways in communities

为小区居民提供环形闭合步道
Provide ring closed footpath for residents

· 利用小区主路人行道，小区围墙沿线空间建设环形健身步道。
· 步道宽度1.2~2m为宜。

· Build ring fitness footpath by using sidewalk of main road and space along the fence of community.
· The footpath width should be 1.2~2m.

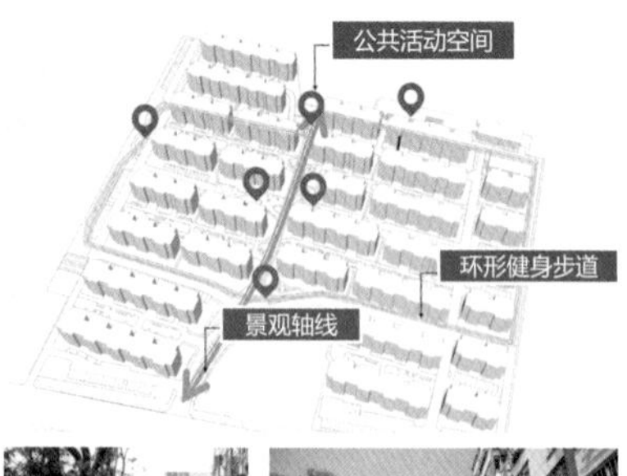

昆山中华园北村环形健身步道： 结合小区道路及围墙沿线空间设置环状健身步道，健身步道以醒目颜色区分，一定间隔设置座椅或亭廊以供休憩。
Ring fitness footpath in Zhonghua Garden North, Kunshan: Ring fitness footpath is built in conjunction with community road and in space along the fence, the footpath is distinguished with conspicuous colors, with seats or kiosks at certain intervals for rest.

图2-7-23 昆山中华园北村环形健身步道 / Circular fitness trail in North Village of Zhonghua garden, Kunshan | 图片来源：上图：昆山中华北村小区改造规划项目；下图：编写组自摄

步道材质、颜色及图标增加识别性
Footpath materials, color and signs for better cognition

· 环状步道以沥青或透水混凝土为主要材质，富有一定弹性。
· 可利用色彩进行区分。
· 步道宜采用无障碍设计原则，避免使用台阶。
· 步道两侧及路面可适当添加表示距离的标识或短句标语。

· The ring footpath shall be mainly made of asphalt or pervious concrete with some elasticity.
· Different colors can be used for distinction.
· The barrier-free design principle should be adopted, to avoid use of steps.
· Signs to indicate distance or short slogans can be added as appropriate on both sides of footpath and pavement.

图2-7-24 健身步道以彩色区分 / Fitness footpath distinguished by different colors | 图片来源：编写组自摄

图2-7-25 结合小区主要道路设置健身步道 / Set up fitness footpath in conjunction with main road | 图片来源：编写组自摄

7.7 打造小区周边美丽街道
Build beautiful streets around communities

优化街道断面，提供充足的人行空间
Optimize street sections to provide sufficient space for pedestrian

· 社区生活性街道可适当压缩车行空间，拓宽人行空间。
· 有条件的可设置设施带，统筹布局行道树、市政设施、休息设施、自行车停车等功能，减少设施布设对人行空间的占用和影响，同时形成与机动车道的分隔。

· In streets for living purpose, vehicle space can be reduced while pedestrian space be widened as appropriate.
· Facilities belts can be set up when conditions permit, and functions such as road trees, municipal facilities, rest facilities and bicycle parking can be arranged as per overall layout, so as to reduce the occupation and influence of facilities on pedestrian space, and also separate it from motor vehicle lanes.

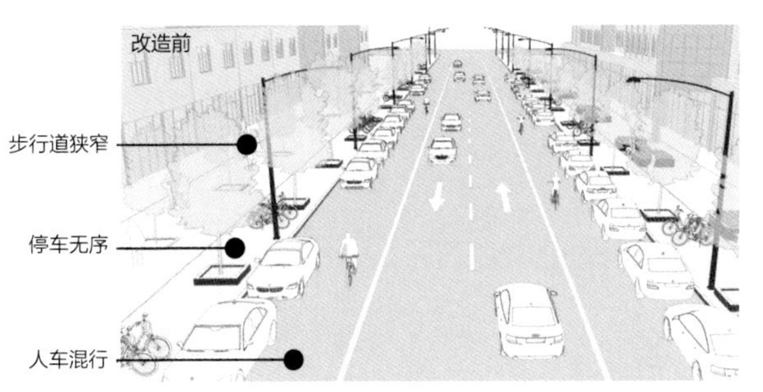

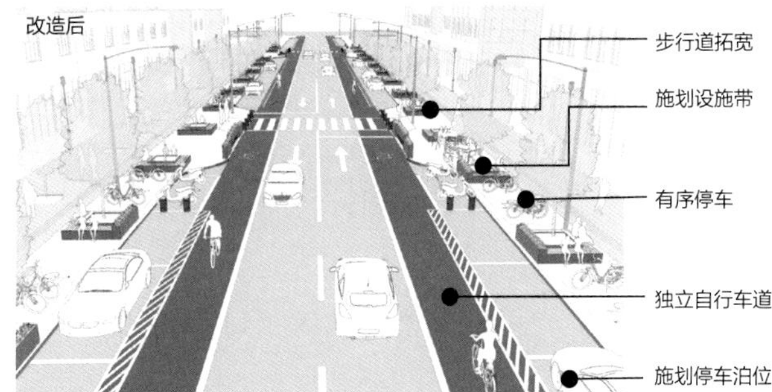

图2-7-26 道路断面优化改造 / Optimization and renovation of road section | 图片来源：嵩山路完整街道改造设计

图2-7-27 拓宽人行道增加店前休闲空间 / Widen the sidewalks for more leisure space in front of store | 图片来源：编写组自摄

图2-7-28 充足的人行空间有利于提供休闲设施 / Sufficient pedestrian space to facilitate providing leisure facilities | 图片来源：编写组自摄

7.7 打造小区周边美丽街道
Build beautiful streets around communities

保障步行空间畅通连续
Ensure well-through and continuous walking space

· 修整人行道铺装，做到平坦、顺畅、防滑，保障步行安全。
· 沿线地块出入口进行抬高式设计，保证步行与盲道连续。

· Repair sidewalk pavement, to make it flat, smooth, non-slippery, to ensure walking safety.
· The entrance and exit of plots along the path are designed with elevation to ensure the continuity of walking and path for the blind.

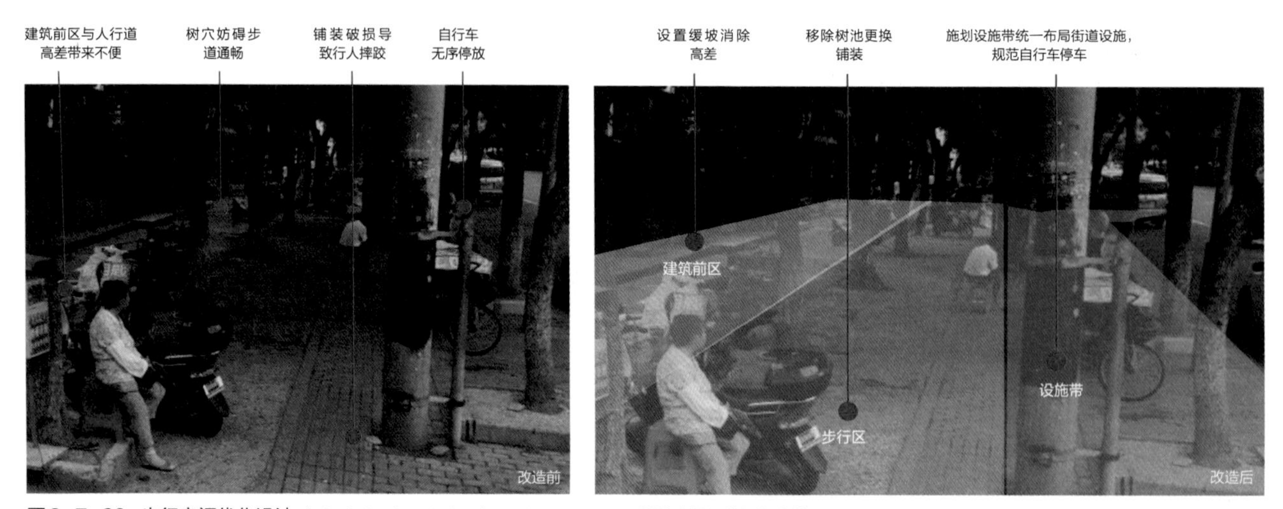

图2-7-29 步行空间优化设计 / Optimization design for walking space | 图片来源：编写组自绘

建筑后退空间与人行道一体化设计
Integrated design for building retreat space and sidewalks

- 通过铺装、色彩等区分人行道和建筑后退空间。
- 建筑后退空间应统一设计，允许沿线商店利用后退空间布设外摆、绿化等设施，丰富沿街活动和景观。
- 处理好建筑后退空间与人行道的高差关系，通过台阶、坡道等连接步行道和店前空间；存在较小高差时，应采用徐缓坡道而非踏步。

- Distinguish sidewalks and building retreat space with pavement and colors.
- Unified design shall be made for building retreat space, and shops along the building shall be allowed to arrange external articles and plantation in the retreat space, to enrich subsequent and landscape along the street.
- Different heights of building retreat space with sidewalks shall be properly treated, and footpath and space in front of stores shall be connected by steps and ramps; for small height difference, gentle ramp shall be used instead of steps.

图2-7-30 建筑后退空间外摆设施 / External facilities in building retreat space | 图片来源：编写组自摄

图2-7-31 通过创意景观分隔空间领域 / Separation of space areas with creative landscape | 图片来源：编写组自摄

图2-7-32 多杆合一集约设置 / Combine posts with different functions | 图片来源：编写组自摄

增加街道休息设施
Add rest facilities along street

- 沿人行道每隔100~125m设置公共座椅，方便老人、儿童歇息。
- 休息设施可以结合设施带布局，也可结合交叉口缘石外拓形成小型休息等待空间。

- Public seats shall be provided at interval of 100~125m along the sidewalks, for the convenience of the elderly and children.
- Rest facilities can be arranged in facilities belt, or curbs at intersections can be extended to form small rest and waiting space

图2-7-33 规范非机动车停车设施 / Formalize parking facilities for non-motorized vehicles | 图片来源：编写组自摄

图2-7-34 路口休息等候设施 / Rest and waiting facilities by road | 图片来源：编写组自摄

图2-7-35 休息座椅与绿化结合设置 / Combination of rest seats with plantation | 图片来源：嵩山路完整街道改造设计

图2-7-36 休息设施结合设施带布置 / Rest facilities arranged in facilities belt | 图片来源：嵩山路完整街道改造设计

7.7 打造小区周边美丽街道
Build beautiful streets around communities

图2-7-37 艺术围墙 / Artistic fence | 图片来源：编写组自摄

美化沿街围墙界面
Beautify fences along street

- 利用栅栏式围墙破墙透绿，使街景与住区相互渗透；运用立体绿化组合的形式美化围墙，使其更具观赏性。
- 通过彩绘、绿植等多元手法增加围墙的艺术气息；增加橱窗等设计手法，使围墙成为文化宣传的窗口。
- Use palisade fence to show the green, for mutual penetration of street landscape with residential area; beautify the fence with the form of 3D green combination, make it more ornamental.
- Color painting, green plants and other methods can be used to make the fence more artistic; add window and adopt other design techniques, to turn the fence into windows of cultural propaganda.

整饰沿街建筑立面
Decorate building facade along streets

- 鼓励建筑底层功能混合，美化橱窗设计。
- 整齐沿街店铺招牌，和而不同。
- 统一整饰建筑外挂空调机架、遮阳设施，做到和谐有序。
- The mix of functions on the ground floor is encouraged to beautify the show window design.
- Line up the shop signs along the street, to be harmonic while different.
- Unify the air conditioner frames and sunshade facilities on the building, to achieve harmony and order.

图2-7-38 围墙透绿 / Show green through the fence | 图片来源：编写组自摄

图2-7-39 统一整饰住宅外立面和沿街商业店招牌 / Rectify the house facade and shop signs along the street on a unified basis | 图片来源：编写组自摄、自绘

提升路面铺装与环境小品的艺术性与趣味性
Make pavement and environmental sketches more artistic and interesting

- 人行道推广生态铺装，采用彩色透水混凝土铺装材料。
- 结合街道氛围，设计符合人体尺度、舒适、具有趣味性的街道家具。
- Ecological pavement shall be popularized for sidewalks, and color pervious concrete pavement materials shall be used.
- Comfortable and interesting street furniture fitting human body shall be designed in conjunction with the street atmosphere.

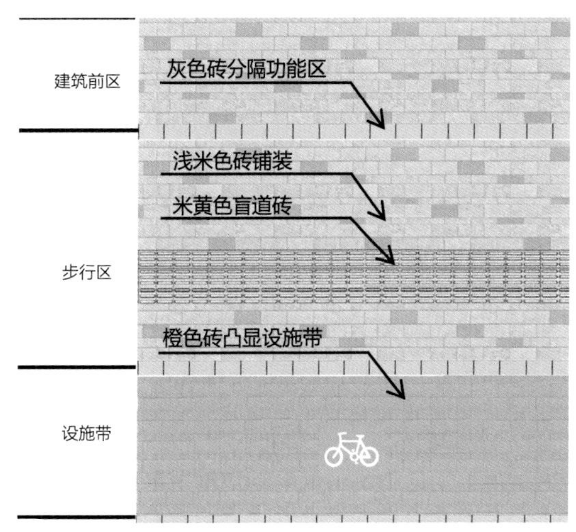

图2-7-40 结合分区的差异化铺装设计 / Differential pavement design also with zonation | 图片来源：编写组自绘

提升沿街绿化景观
Upgrade green landscape along street

- 行道树宜选用当地树种，体现本地化特征。
- 结合绿化带、围墙设施搭配花卉品种，实现见绿见花、四季缤纷。
- 树木绿化可与街道家具并置，将生态和使用功能有机结合。
- 沿街结合建筑后退组织多种开放空间，如广场、口袋公园等，提供多样化活动场所。
- Local tree species should be selected for the street, to reflect the characteristics of localization.
- Green belt and fence facilities should be decorated with different flower species, so that they are green with flowers and blooming in all seasons.
- Trees can be arranged with street furniture to combine ecology and use function organically.
- A variety of open spaces, such as squares and pocket parks, can be organized along the street and building retreat space, to provide a variety of activity venues.

图2-7-41 绿化景观与休息设施、街道家具整体设计 / Overall design of green landscape, rest facilities, and street furniture | 图片来源：嵩山路完整街道改造设计

图2-7-42 趣味休闲座椅 / Interesting rest seats | 图片来源：编写组自摄

7.8 多方式连通形成社区绿道
Realize community green roads in multiple ways

图2-7-43 绿道沿线布设健身设施 / Fitness facilities along green road | 图片来源：编写组自摄

满足居民就近散步、健身的需要
Meet the needs of walking and fitness of residents nearby

- 提供成环的步道系统，满足周边居民不同散步、健身距离的需求。
- 加强与上一级（城区级）绿道网络的衔接。
- Provide a ring footpath system, to meet different walking and fitness distance demand of nearby residents.
- Strengthen linking with higher level (urban level) green road network

充分利用消极空间，让绿色看得见也走得近
Make full use of idle space, so that green is visible and accessible

- 利用滨水绿带空间、用地后退道路红线空间、快速路、轨道沿线的防护绿带建设社区绿道。
- 空间有限的情况下，可以借用道路人行道空间。
- 连接形成闭环的社区级绿道网，宽度不小于1.2m，提升沿线绿化景观，统一铺装样式，加强空间标识性。
- Build community green belt in waterfront green space, the red line space of the receding land from road, the expressway and the protective green belt along the track lines.
- The space of sidewalks of roads can be used when space is limited.
- Form community level green road network in closed loop with width no less than 1.2m, to improve green landscape along the line, unify pavement style and strengthen space labeling.

图2-7-44 结合滨河空间建设绿道 / Build green ways in riverside space | 图片来源：编写组自摄

图2-7-45 借用道路人行空间建设绿道 / Build green ways in walking space of road | 图片来源：编写组自摄

绿道应尽可能地串联服务设施和绿地公园
Green roads shall connect service facilities and green land and parks as much as possible

绿道沿线应提供人性化的服务设施
Humanized service facilities should be provided along the greenway

- 安全设施：太阳能照明灯具、监控摄像头。
- 休息设施：座椅、遮阳避雨的亭廊花架。
- 市政设施：分类垃圾桶、移动公厕、宠物厕所。

- Safety facilities: solar lighting fixtures, surveillance cameras.
- Rest facilities: seats, sun shade and rain pavilion gallery flower racks.
- Municipal facilities: waste classification bins, mobile public toilets, pet toilets.

图2-7-46 展示特色景观 / Demonstrate characteristic landscape | 图片来源：编写组自摄

展示独特主题景观或社区文化
Display a unique thematic landscape or community culture

- 不同路段可以结合沿线资源，形成不同的主题特色。
- 设置能反映社区文化的创意小品雕塑，营造独特景观。

- Different sections form different theme characteristics by combining with resources along the route.
- Set up creative sketch sculptures that can reflect the culture of the community to create a unique landscape.

图2-7-47 绿道连接社区公园与街头绿地 / Green road connecting community part and street green land | 图片来源：编写组自摄

Construction Guidance for
Renovation of Old Communities in
Jiangsu

江苏老旧小区改造建设导引

8

提升绿化环境景观

Enhance the Environment Landscape

8.1 多方式提高小区绿视率
Raise community green looking ratio in multiple ways

完善小区绿地系统，鼓励立体绿化
Complete green system in community and encourage 3D plantation

- 依据现状条件完善点、线、面相结合的绿地系统。
- 提升公共绿地景观：合理划分功能区域，形成疏密有致、方便进入与观赏的绿地景观。
- 修整院落宅旁绿地：表土裸露区域合理补栽花草树木，房屋南侧不宜种植高大乔木，避免遮挡阳光。
- 道路沿线绿地：注重对原有行道树保护和利用，沿线绿化设计应发挥雨水管理作用。
- 鼓励垂直绿化：结合屋顶、围墙、廊架、车棚等进行复合设计，既可提升绿化视觉景观，也可发挥吸收外界噪声、净化空气的生态作用。

- Complete the green space system combining points, lines and area according to current conditions.
- Upgrade the landscape of public green space: rationally divide the functional areas to form green space landscape with different density, easy to access and enjoyable.
- Rectify the green space beside courtyard houses: reasonably plant flowers and trees in the exposed area of the topsoil, and it is not appropriate to plant tall trees in the south side of houses, to avoid blocking the sunshine.
- Green space along road: pay attention to the protection and utilization of existing street trees, and the greening design along the road should pay attention to use of rainwater.
- Encourage vertical greening: carry out composite design for roof, fence, corridor frame, shed and so on, to improve the visual landscape of greening, and also play the ecological role of absorbing external noise and purifying air.

梳理原有乔灌草植物布局，形成疏密有致、方便进入和观赏的绿地景观。
选取易生长、易维护的本土植物。
适当增加开花类乔灌木和色叶植物，丰富季相景观。
增设健身步道。
Re-arrange original layout of trees, shrubs and grasses, to form green space landscape with different density, easy to access and enjoyable.
Select native plants that are easy to grow and maintain.
Add flowering trees, shrubs and color leave plants, to enrich the seasonal landscape.
Add fitness footpath.

图2-8-1 小区公共绿地改造设计 / Design for public green land renovation in community | 图片来源：编写组自摄

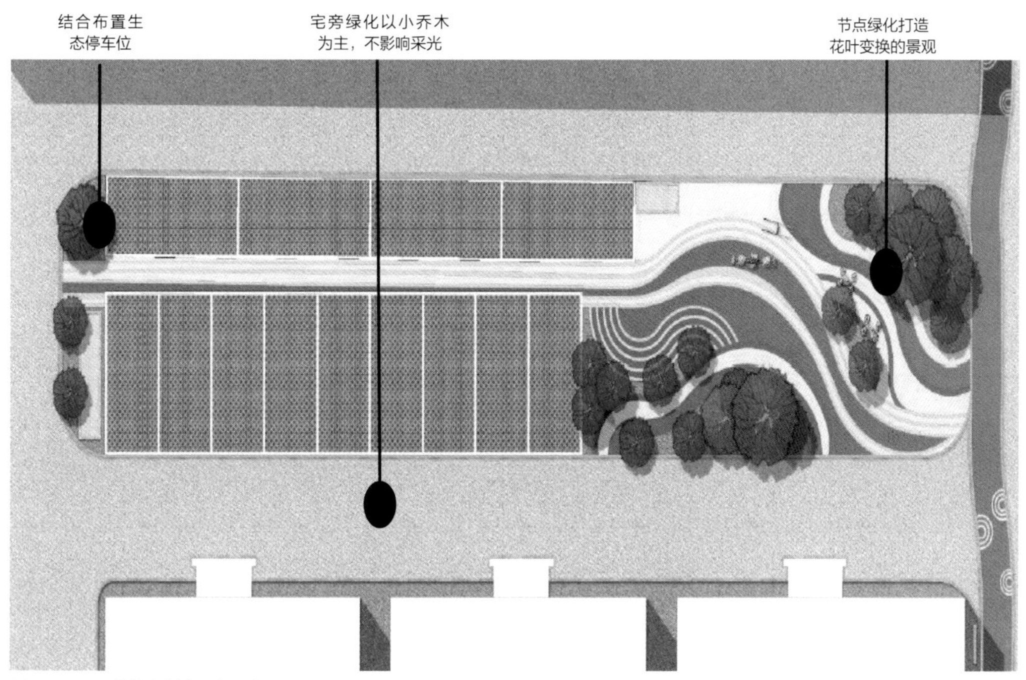

图2-8-2 院落绿地提升设计 / Upgrading design for courtyard green land | 图片来源：琼花新村老旧小区改造项目成果

图2-8-3 院落绿地提升设计 / Upgrading design for courtyard green land | 图片来源：琼花新村老旧小区改造项目成果

8.1 多方式提高小区绿视率
Raise community green looking ratio in multiple ways

植物配置应兼顾生态、景观、功能、管养要求
Plant configuration should take into account the ecology, landscape, function and management requirements

图2-8-4 活动场地高大乔木与休闲设施结合布局 / Combination of high trees and leisure facilities in activity venues | 图片来源：编写组自摄

- 适当体现四季景观：乔、灌、花、草合理配置，多选用香花色叶类、易成活、轻管养、效果好、造价低的品种。
- 适应不同区域功能需要：小区主路两侧、活动场地周边宜栽植高大乔木，满足夏季遮阴需求；儿童游戏、老人活动场地周围不得选用有毒、有针刺、有臭味、多飞絮的植物。
- 体现乡土特色：植物品种宜选用乡土适生植物，体现地方特色及便于养护。
- 适当增加开花及色叶植物，增添小区色彩。

- Reflect the landscape of four seasons appropriately: trees, shrubs, flowers and grass shall be reasonably allocated, more varieties with fragrant flowers and color leaves, easy to grow requiring little care, with good effect and low cost shall be planted.
- Meet the functional needs of different areas: tall trees should be planted on both sides of main roads and around the activity site, to meet the demand for shade in summer; poisonous, needle-pricking, smelly and fluffy catkins plants must not be used around playground for children and the elderly.
- Reflect native characteristics: suitable native plants should be selected as plant varieties, to reflect local characteristics and facilitate maintenance.
- Plant more flowering and color leaf plant appropriately, to enrich colors of the community.

图2-8-5 小区集中绿地改造为雨水花园 / Centralized green land in community transformed into rainwater garden | 图片来源：编写组自摄

共创共享共维，发动居民参与设计和维护

Joint creation, share and maintenance, mobilize residents to participate in design and maintenance

- 鼓励组织居民共同参与，从设计、实施、维护，到持续更新，进行全生命周期的管理，打造被居民认可和喜爱的景观。
- 提高绿地空间景观的参与性与互动性，采用丰富的互动性良好的植物，让居民体验种植、采摘、食用的喜悦，组织趣味活动，打造小区自然课堂。

- It is encouraged to organize residents to participate in the whole life cycle management from design, implementation and maintenance to continuous renewal, so as to create a landscape recognized and loved by residents.
- Improve the participation and interaction of green space landscape, rich and well-interactive plants can be used to let residents experience the joy of planting, picking and eating, organize interesting activities, and create a natural classroom of the community.

上海虹旭小区居民共建“生境花园”：
利用小区南侧一块拆违后的边角地改造为“生境花园”，成为一处生态科普休闲区；优选50多种乔灌草种类，让居民体验农事，感受收获；引入海绵理念，采用透水铺装，凉亭平台设有雨水收集桶，帮助浇灌果蔬。

“Habitat Garden” jointly built by residents of Hongxu Community, Shanghai:
A piece of edge land with illegal building demolished in the south of the community was transformed into the “habitat garden”, as an ecological popular science leisure place; more than 50 species of trees, shrubs and grasses were selected to allow residents to experience farming and harvest; the sponge concept was introduced and pervious paving was adopted, the pavilion platform is provided with rainwater collection bucket for irrigating the fruits and vegetables.

8.2 提升入口景观
Enhance entrance landscape

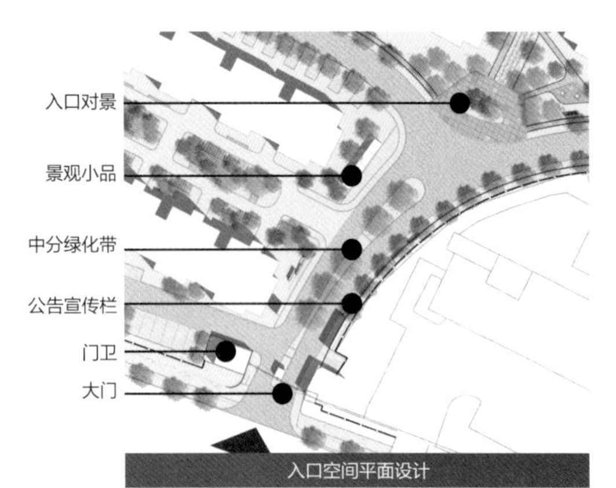

入口空间平面设计

入口中分绿化带景观

具有江南特色的大门造型

图2-8-6 昆山中华园北村入口空间整体改造 / Overall renovation of entrance space in Zhonghua Garden North, Kunshan | 图片来源：昆山中华园北村老旧小区改造项目成果

▶ 整体改造入口空间
Overall renovation of entrance space

- 结合小区入口功能完善和交通流线优化，整合入口空间要素，包括大门、门卫、岗亭、绿化等，进行一体化设计。风格应与小区整体风貌相协调，形成具有识别性的入口标志。
- 提升入口绿化景观，包括入口通道两侧、中分绿带、道路对景等绿化空间，加强绿化层次，适当配置四季花卉，提升观赏性。

- Integrate the entrance space elements, including gate, gate guard, sentry box and plantation, in combination with the improvement of the entrance function and the optimization of traffic flow lines. The style should be coordinated with the overall style of the residential area to form a recognizable entrance sign.
- Enhance the green landscape at the entrance, including the green space on both sides of the entrance passage, the green belt in the middle and the landscape on the road, strengthen the green level, and properly configure flowers of four seasons to enhance the ornamental value.

通过多样化元素营造归家氛围
Create an atmosphere of returning home with diversified elements

- 通过增设入口步道、休闲座椅、风雨长廊等休憩设施，提升居民归家的舒适性。
- 贴合小区主题元素进行装饰设计，营造特色化的入口空间氛围。

- Make residents more comfortable when back home by adding rest facilities such as entrance footpath, leisure seats and shelter corridor.
- Make decoration design on theme elements of the residential area to create a characteristic atmosphere of the entrance space.

图2-8-7 昆山枫景苑小区入口道路现状 / Current situation of entrance road in fengjingyuan community of Kunshan | 图片来源：昆山枫景苑A区老旧小区改造项目成果

图2-8-8 昆山枫景苑小区入口归家氛围营造 / Create a home return atmosphere at entrance of Fengjing Garden, Kunshan | 图片来源：昆山枫景苑A区老旧小区改造项目成果

8.3 提升围墙艺术性和功能性
Improve artistic and functional performance of fences

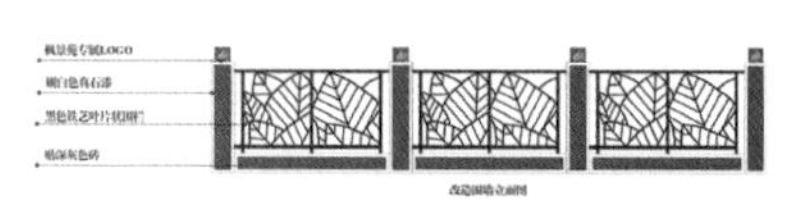

图2-8-9 围墙高度适宜 / Suitable height of fence | 图片来源：枫景苑A区改造设计项目

图2-8-10 围墙形式、色调与周边环境相协调 / Fence in harmony with the surrounding environment in form and tone | 图片来源：枫景苑A区改造设计项目

▶ 围墙应与小区风貌和周边环境协调
The fence shall be harmonic with community style and surrounding environment

· 高度：体现宜人尺度，一般小区围墙高度宜为2.1~2.4m，最小高度不宜低于1.8m。

· 形式：简洁雅致，与周边环境相协调。

· 材质：注重经济、地方特色和亲切感。

· 增加细节设计，如入口处围墙结合小区LOGO、景观绿化等组合设计，提升精致感。

· Height: it shall be in a pleasant scale. Generally, the height of the community fence should be 2.1~2.4m, and the minimum height should not be less than 1.8m.

· Form: concise and elegant, in harmony with the surrounding environment.

· Material: it shall be economic, with local features and sense of intimacy.

· More details shall be added, such as the community LOGO on fence at entrance, combination of landscape and plantation, to make it more refined.

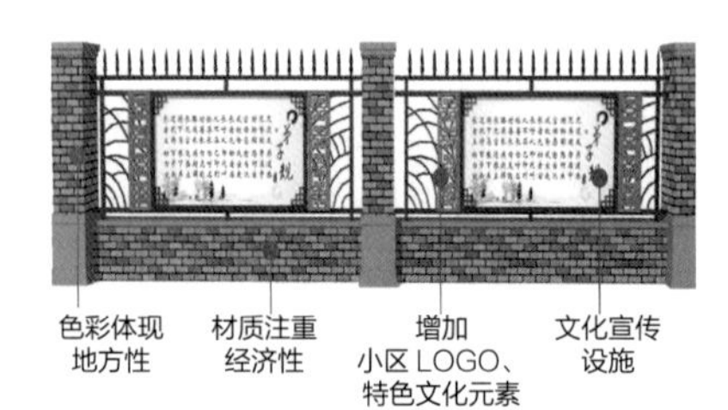

图2-8-11 注重经济、地方特色和亲切感 / Be economic, with local features and sense of intimacy | 图片来源：昆山蓬曦园老旧小区改造项目成果

▶ 围墙功能复合利用
Composite utilization of fence functions

· 鼓励围墙复合功能利用，嵌入党建宣传、文化展示等功能，丰富围墙景观。

· 结合绿化、照明、休闲座椅等景观元素整体设计，实现“一墙多用”，为城市提供更多方便。

· It is encouraged to develop composite functions of fence, to add functions such as Party building publicity and culture display, to enrich the fence landscape.

· Realize “diversified use of fence” with integrated design for plantation, lighting and leisure seats, to provide more convenience for the city.

8.4 提升住区环境特色与文化氛围
Improve environmental features and cultural atmosphere in living quarters

合理运用地方建筑文化元素
Make rational use of local architectural culture elements

- 建筑风貌整治应符合上位规划关于所在分区的风貌、色彩等控制要求。
- 挖掘地域传统建筑文化，提炼地方建筑元素和色彩符号，结合外立面设计合理运用，彰显地方建筑特色。
- 注重檐口、线脚、单元门头、设施构件等细部设计，提升细节感和精致感。

- Architectural style renovation shall comply with control requirements of upper level planning for styles and colors of the zone.
- Tap traditional architectural culture of the area, extract the local architectural elements and color symbols, make rational use in combination with the facade design, to highlight the local architectural features.
- Pay attention to detail design of cornice, architrave, unit door head, facility components, etc., to enhance the sense of detail and delicacy.

图2-8-12 细节设计体现地方建筑文化特色 / Detail design reflects the local architecture culture feature | 图片来源：昆山枫景苑A区老旧小区改造项目成果

枫叶主题小区LOGO设计 / LOGO design for maple leave theme community

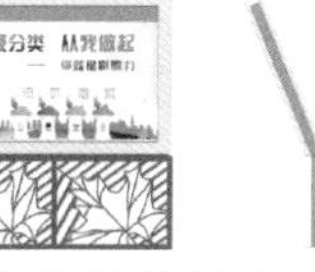

公告宣传栏 / Bulletin publicity column

楼栋指示牌 / Building signboard

小区总览图 / Community overview

警示提示牌 / Warning and prompt plate

图2-8-13 昆山枫景苑小区以“枫”为主题进行整体环境设计 / Overall environment design with “maple” as the theme for Fengjing Garden, Kunshan | 图片来源：昆山枫景苑A区老旧小区改造项目成果

注重环境要素的系列化设计
Pay attention to series design of environmental elements

- 环境小品：鼓励运用地方材料，选择乡土树种，体现地方景观特色。
- 标识系统：对小区导览图、公告宣传栏、路标指示牌、楼栋标识牌等进行系列化设计，体现整体感和文化性，同时应与建筑、环境整体风格呼应。
- 通过建筑、环境小品、标识系统等的整体重塑，强化小区整体风貌特色，并以此推动地区风貌环境的不断优化。

- Environmental sketch: encourage the use of local materials, choose native tree species, to reflect the local landscape characteristics.
- Signage system: series design shall be made for the community guide map, bulletin board, road sign board, and building sign board, to reflect the sense of integrity and culture, and also echo with the overall style of the buildings and environment.
- Through the overall remodeling of buildings, environmental sketches, signage system, the overall style and features of the community can be enhanced, to promote the continuous optimization of features and environment of the area.

8.4 提升住区环境特色与文化氛围
Improve environmental features and cultural atmosphere in living quarters

保护利用特色资源
Protect and use featured resources

- 深入挖掘所在地区的人文内涵，保护小区中的历史遗存、古树名木及后续资源、河塘水体等特色要素，结合公共设施、公共空间予以合理利用，保护风貌延续记忆，使其成为小区独特的景观。
- 结合小区的历史特色，赋予公共空间一定的文化主题，结合人物、事件、场所等创作，强化特色表达。

- The cultural connotation of areas shall be deeply explored to protect the historical remains, ancient and famous trees, follow-up resources, river ponds and water bodies and other feature elements in the community, and to make rational use of them in combination with public facilities and public space, so as to protect the style and continue the memory and make them the unique landscape of the community.
- Combined with the historical features of the community, the public space is endowed with a certain cultural theme, to strengthen the expression of features with the creation of figures, events and venues.

重庆嘉陵桥西村通过保护利用特色资源营造独特社区环境：
保护小区中的市级历史文化遗存：马歇尔故居和鲜英故居，结合公共空间予以合理利用，形成主要的文化线路以及小区独特的景观。
Create a unique community environment by protecting and utilizing featured resources in Jialingqiao West Village, Chongqing:
Protect the municipal historical and cultural relics in the community: former Residence of Marshall and Former Residence of Xian Ying, to be rationally utilized in combination with public space to form the main cultural routes and the unique landscape of the community.

鼓励居民自主的文化表达
Residents are encouraged to express their own culture

- 引导社区居民群众发动自身力量，进行文化创意表达，如楼栋美化、围墙彩绘、园艺展示等，创造小区个性特色。
- 鼓励自发的“微改造”行动，结合小区公共空间增设公共艺术作品、活动设施，形成独特景观。
- 挖掘社区传统文化，通过多种形式加以传承，弘扬社区文化。

- Guide the community residents to mobilize their own force for cultural and creative expression, such as building beautification, wall painting, horticulture display, etc., to create the personalized characteristics of the community.
- Encourage the spontaneous “micro-renovation” action, to add public art works, activity facilities in the community public space, to form a unique landscape.
- Tap the community traditional culture, and carry forward the community culture through inheriting in various forms.

重庆锦天佳园打造“家文化”主题小区：
以“家文化”为主题，打造家风主题小区。
用“家和”“家孝”“家廉”“家训”“家梦”为小区道路、中心花园命名。
在社区各个角落打造文化活动阵地。
Create “family culture” theme community in Jintian Jiayuan, Chongqing:
Create family style theme community with “family culture” as the theme.
Name the community roads and central garden as “family harmony” “family piety” “family honesty” “family motto” and “family dream”.
Turn all corners of the community as cultural activity venues.

Construction Guidance for

Renovation of Old Communities in Jiangsu

江苏老旧小区改造建设导引

9

规范物业管理

Standardize Property Management

- 9.1 配置物业管理用房
  Arrange premises for property management
- 9.2 政府托底的物业管理
  Property management backed up by the government
- 9.3 引导小区成立业委会等自治组织
  Guide the establishment of autonomous organizations such as owners' committee in communities
- 9.4 鼓励小区引入市场化专业物业管理服务
  Encourage communities to introduce marketized professional property management services
- 9.5 有条件的小区可以提供物管人员住宿
  Provide accommodation for property management personnel in communities with ready conditions
- 9.6 划零为整，统一物业管理
  Unify diversified property management

9.1 配置物业管理用房
Arrange premises for property management

《江苏省物业管理条例》第三十五条：
按照不低于地上地下总建筑面积千分之四的比例配置物业服务用房，低于一百平方米的按照一百平方米配置。其中，用于业主委员会议事活动用房的，应当按照配置物业服务用房的比例合理确定，一般按照建筑面积二十至四十平方米配置。
Article 35 of *Property Management Regulation of Jiangsu Province*:
Property service premises shall be arranged in the proportion of no less than 4‰ of the total floor area above ground and underground, and those less than 100 square meters shall be allocated as 100 square meters. The room for meeting or discussion activity of owner committee shall be determined at a rational ratio of the property service premises, normally with floor area of 20 to 40 square meters.

图2-9-1 社区党群服务中心 / Community Party and mass service center | 图片来源：编写组自摄

扬州荷花池小区：
扬州荷花池小区结合社区用房改造物业管理用房，不仅满足功能，建筑风格上也与小区其他建筑统一；内部除了承载物业管理办公室的作用，还兼顾社区宣传展示、社区活动开展等功能。
Hehuachi Community, Yangzhou:
The community has renovated premise in the community for property management purpose, not only meeting the functions, but also in unified architectural style with other buildings in the community; the rooms, in addition to being used as property management office, have the functions of community publicity and community activities.

图2-9-2 扬州荷花池小区物业管理用房改造 / Renovation of premises for property management in Hehuachi Community, Yangzhou | 图片来源：编写组自摄

图2-9-3 物管用房的展示墙 / Exhibition wall of property management premises | 图片来源：编写组自摄

▶ 物业管理用房应满足物业管理需要
Property management premises shall meet the management need

- 老旧小区改造应完善物业管理用房，配置标准参考《江苏省物业管理条例》或地方相关标准。
- 有条件的小区可以增设小区议事厅、党群服务站等社区服务设施，方便居民开展活动。
- In renovation of old communities, property management premises shall be completed according to the *Property Management Regulation of Jiangsu Province* or relevant local standard.
- Where conditions permit, community meeting rooms, Party and mass service stations and other community service facilities may be added to facilitate the activities of residents.

▶ 利用小区现状公共用房、公共租赁等设施进行改造
Make use of existing public premises and public leased facilities for renovation

- 物业管理用房根据小区场地条件，采用集中、分散相结合的方式完善。
- 结合小区现有公共用房改造或扩建增设物业管理用房，完善门卫、岗亭等物业管理设施。
- 在符合法律法规的前提下，租用住户住宅底层或架空层进行改造。
- 在不影响周边住户环境的前提下，利用小区空地建设物业管理用房，需处理好与相邻环境的协调。
- Property management premises shall be completed according to the available conditions in both centralized and distributed forms.
- Existing public houses can be renovated or expanded for property management purpose, to complete property management facilities such as gate guard and kiosk.
- The ground floor or elevated layer of resident buildings can be leased for renovation while conforming to laws and regulations.
- Property management house can be built on space in the community without affecting the environment of surrounding households, and good coordination is required with neighboring environment.

9.2 政府托底的物业管理
Property management backed up by the government

委托国有平台打包管理
Entrust a state-owned platform for package management

- 政府对于老旧小区应有托底服务，探索成立国有平台，打包管理老旧小区。
- 国有平台接手初期，可以根据小区条件，提供基本服务，包括保安、卫生保洁、绿化养护等。
- 国有平台在运营一段时间后，可以根据小区实际情况，增加多元化、个性化服务，通过良好的服务，逐步培养居民“花钱买服务”的意识。

- The government should provide support services for old communities, explore establishing state-owned platform for package management of old communities.
- In the early period of takeover, the state-owned platform can provide basic services, including security guard, cleaning, greening maintenance, etc., according to the conditions of the community.
- After a period of operation, the state-owned platform can increase diversified and personalized services according to the actual situation of the community, and gradually cultivate residents' awareness of “paying for services” by providing high quality services.

图2-9-4 姚坊门物业管理公司 / Yaofangmen Property Management Company | 图片来源：编写组自摄

图2-9-5 红色物业 / Red property management | 图片来源：编写组自摄

南京尧化街道物业：
南京尧化街道积极推进“红色物业”建设，充分整合基层“红色资源”，以糅合“社区党组织+小区物业党员阵地+党员志愿者”的新模式为支点，提升物业管理与服务水平。

Property management of Yaohua Subdistrict, Nanjing:
Nanjing Yaohua Subdistrict actively promotes the construction of “red property management”, fully integrates the “red resources” at the primary level, and takes the new model of “community Party organization + community property Party member positions + Party member volunteers” as the fulcrum to improve the property management and service level.

图2-9-6 社区党组织与两新党组织结对共建 / Joint construction of community Party organization with Party organizations in new social and new economic entities | 图片来源：姚坊门物业公司提供

红色物业扎根基层
Red property management taking room in the primary level

- 加强现有物业公司的基层党建工作。
- 选聘党员参与物业服务管理工作，把物业服务纳入党的工作队伍，推动基层社会治理。

- Strengthen the primary level Party building work in existing property companies.
- Select Party members to participate in property service and management work, to incorporate the property service into the working team of the Party, to push forward primary level social governance.

图2-9-7 姚坊门物业在疫情期间冲锋在前 / Yaofangmen property management personnel act as vanguard during the epidemic period | 图片来源：姚坊门物业公司提供

9.3 引导小区成立业委会等自治组织
Guide the establishment of autonomous organizations such as owners' committee in communities

引导成立业委会依法履行职责
Guide to set up owners' committee to exercise duties pursuant to law

- 街道、社区及其党委要发挥引领作用，引入物业企业的同时推动小区成立业委会。
- 业委会需要发挥应有的作用，凝聚业主力量监督物业公司履行物业服务合同，形成议事和执行机构，处理业主集体事务，制定小区居民行为守则等。
- 加强社区党组织、社区居委会对业主委员会的指导和监督，将小区业委会组建和换届工作纳入基层党建工作的重点内容。逐步提高业委会中的党员比例，原则上不少于半数。

- Subdistricts, communities and their Party committees should play a leading role in introducing property enterprises while promoting the establishment of community owners' committees.
- The owners' committee shall play its due role, concentrate the owners' power to supervise the property company to perform the property service contract, form a discussion and execution organization, deal with the owners' collective affairs, and formulate the code of conduct for residents of the community.
- The guidance and supervision by Party organizations and community neighborhood committees over owners' committees shall be strengthened, and the formation and change of community owners' committees shall be taken as key content of primary-level Party building work. The proportion of CPC members in the owners' committee shall be raised gradually, to be no less than half in principle.

定期召开业主大会
Call owners' conference periodically

- 按照业主大会议事规则，业主委员会应定期组织召开业主大会。
- 业主大会会议可以采用集体讨论的形式，也可以采用书面征求意见的形式；但应当由物业管理区域内专有部分占建筑物总面积过半数的业主且占总人数过半数的业主参加。

- The owners' committee shall call owners' conference periodically according to the rule of procedure of owner's conference.
- A owners' conference may be held in the form of collective discussion or written solicitation for opinions; however, it shall be attended by owners with exclusive parts accounting for more than half of the total area of the buildings of the owned part of the property management area and more than half of the total number of owners.

暂不具备成立业委会条件的，可由街道组织成立物业管理委员会

A property management committee can be set up by the subdistrict where conditions are tentatively not ready for setting up a owners' committee

- 物业管理委员会由街道办事处（乡镇人民政府）代表、社区居民委员会代表、辖区派出所代表、建设单位代表和业主代表组成，体现业主、物业服务企业和属地政府三方对小区的共建共治共享。
- 物业管理委员会为公益性组织，无偿为小区业主服务，代行业主大会和业主委员会职责。

- The property management committee shall be formed by representatives of the subdistrict office (township people's government), the community neighborhood committee, the area police station, the construction entity and the owners, to reflect the joint construction, governance and sharing of the community by the owners, property service enterprise and the local government.
- The property management committee is a public welfare organization, serving the community owners at no charge, and exercising the duties of the owners' conference and owners' committee.

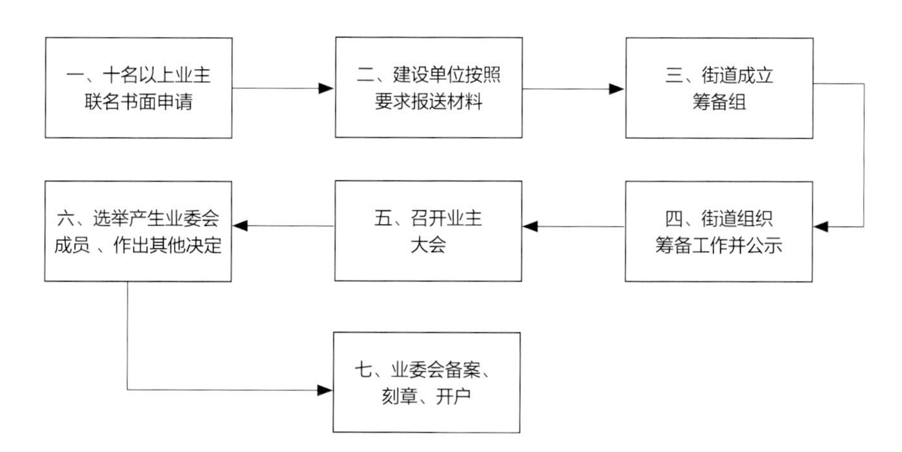

图2-9-8 南京市小区召开首次业主（代表）大会/业主代表大会的流程图 / Flow chart of the first owner (representative) conference/ owner congress in a community in Nanjing | 图片来源：编写组自绘

9.4 鼓励小区引入市场化专业物业管理服务
Encourage communities to introduce marketized professional property management services

及时引入物业服务企业开展规范化服务
Promptly introduce property service enterprise for standardized service

· 物业是为小区业主提供对房屋及配套的设施设备和相关场地进行维修、养护、管理，维护物业管理区域内的环境卫生和相关秩序等活动的服务企业。
· 鼓励小区实施改造前即引入适合的市场化专业物业管理服务，支持物业服务企业全程参与改造过程。
· The property management is a service enterprise that provides the owners of the residential area with repair, maintenance and management of houses and associated facilities and equipment and relevant sites, and maintains the environmental sanitation and relevant order in the property management area.
· Encourage communities to introduce suitable marketized professional property management services before implementing renovation, and support property service enterprise to take part in the whole course of renovation.

图2-9-9 物业管理服务 / Property management service | 图片来源：银城生活服务有限公司提供

银城物业：
银城物业以两大业务线（即物业管理服务和生活社区增值服务）组成一个综合服务平台。通过该两大业务线达成的协同效益，有助银城物业多元化发展收益来源，同时通过向客户提供互补的服务，巩固物业与客户的关系。
Yincheng Property:
Yincheng Property formed a comprehensive service platform with two service lines (namely property management services and value-added services for living communities). The synergic effect achieved by these two lines help to diversify the sources of revenue for Yincheng Property while consolidating the relationship between the Property and its customers by providing complementary services to their customers.

鼓励物业企业提供更加多元化、个性化的服务，不断提高物业管理水平
Encourage property enterprises to provide more diversified and personalized services to continuously improve the level of property management

· 随着需求的增长，物业公司可以探索提供养老医疗、社区资产管理、智慧服务等高品质服务。
· 开发专门的手机APP为业主提供便利服务，并可以通过手机APP促进邻里交流，方便业委会监督。
· 街道办事处对物业管理和后期运营做好监督工作，确保小区物业服务有序。
· As demand grows, property companies can explore providing high-quality services such as old-age care, community asset management and smart services.
· Develop special mobile phone APP to provide convenient services for owners, and promote neighborhood communication via mobile phone APP to facilitate supervision by the owners' committee.
· The subdistrict office shall supervise the property management and subsequent operation to ensure good order in the property service in the community.

图2-9-10 银城物业两大主线业务之一——生活社区增值服务 / One of the two main service lines of Yincheng—— value-added services in living community | 图片来源：银城生活服务有限公司提供

图2-9-11 银城物业两大主线业务之二——物业管理服务 / The second of the two main service lines of Yincheng —— property management service | 图片来源：银城生活服务有限公司提供

9.5 有条件的小区可以提供物管人员住宿
Provide accommodation for property management personnel in communities with ready conditions

物业管理人员需要就近入住
Property management personnel need nearby accommodation

- 物业管理需要提供24小时服务，应创造条件给物管人员提供就近的、可负担的、健康的居住环境，保障物管人员身心健康，提供完善高效的服务。
- The property management provides 24-hour service. So conditions should be created to provide the property management personnel with a nearby, affordable and healthy living environment, protect their physical and mental health, to enable them to provide perfect and efficient services.

如何在小区内安排物管人员宿舍
How to arrange dormitory for property management personnel in a community

- 在符合法律法规的前提下，通过租用住宅底层改造，或利用小区空地建设，或整合小区公共建筑资源改造或扩建等方式，提供物管人员宿舍。
- 新建、扩建设施产权归社区所有，由物业经营，经营收入用于提升服务。
- On condition of conforming to laws and regulations, the dormitory for property management personnel can be provided by renting the ground floor of residence, building houses on the open space of the community, or integrating the public building resources of the community for reconstruction or expansion.
- The ownership of new and expanded facilities belongs to the community and is operated by the property company. The operating income will be used to improve the service.

专业、集中提供物业管理人员住宿
Provide dedicated and centralized accommodation for property management personnel

- 小区内无法解决物管人员住宿的，可以集中化、专业化提供物业管理人员住宿。
- 有条件的物业公司可以单独选址、单独建设，从而解决物业管理工作人员的后顾之忧；也可以与其他企业进行合作，共同建设企业员工用房，分摊成本。
- Dedicated and centralized accommodation can be provided for property management personnel if such accommodation cannot be provided in communities.
- Property companies with ready conditions can build separate facilities on selected site, to solve the after worries of their personnel, and they can also cooperate with other enterprises to jointly build dormitory for employees at apportioned cost.

北京安益星物业：
北京安益星物业管理有限公司，作为北京的职住一体蓝领社区即将正式开始运营。旨在为各类企业集中解决职工居住空间问题。园区内商业配套齐全，包括超市、饭店、面包房、咖啡厅、篮球场、羽毛球场、绿荫大道等。多种房型（8/6/4人间）满足各种需求（长/短租），企业定制，宿舍托管。

Beijing Anyixing Property:
Beijing Anyixing Property Management Co., Ltd., as a blue-collar community integrating work and housing, will officially start operation soon. It is aimed to providing centralized accommodations for employees of various enterprises.
The community has a complete set of commercial facilities, including supermarkets, restaurants, bakeries, cafes, basketball courts, badminton courts, boulevard and so on.
A variety of rooms (rooms for 8/6/4 people) to meet various needs (long/short rent), customized by enterprises, the dormitory is managed under trusteeship.

9.6 划零为整，统一物业管理
Unify diversified property management

整合零散老旧小区，统一物业管理
Scattered old communities will be integrated for unified property management

- 对于分散式、规模小的老旧小区，可以适当组合，街道社区帮助统一引进物业管理，降低物管成本。
- 物管机构可以针对各小区实际情况，开展针对性的菜单式服务，满足不同服务需求。

- Scattered small old living quarters can be combined as appropriate, and the subdistrict and community can help introduce unified property management, to reduce the management cost.
- Property management institutions can offer targeted menu-based service according to their actual conditions, to meet different demands.

图2-9-12　零散小区增设道闸封闭管理 / Add road gates for scattered communities for enclosed management | 图片来源：编写组自摄

图2-9-13　南京宁海路一带零散小区分布图 / Distribution map of scattered living quarters along Ninghai Road in Nanjing | 图片来源：编写组自绘

南京宁海路一带零散小区统一引进物业管理：
宁海路一带零散小区通过设置出入口道闸，将原有分散的宁夏路23号小区、宁夏路13号院、琅琊新村、西康路35号小区形成一个物业区域，引入南京银城物业管理有限公司进行统一管理，物业服务涵盖保洁、秩序维护、绿化养护、工程维修等。

Unified property management for scattered living quarters along Ninghai Road, Nanjing:
Entrance and exit gates were set up for scattered living quarters, to form one property zone for the four quarters at 23 Ningxia Road, 13 Ningxia Road, Langya New Village and 35 Xikang Road, and Nanjing Yincheng Property Management Co., Ltd. was engaged for unified management, covering cleaning, order maintenance, plantation and caring, and engineering repair.

培育街区认同感和责任感
Foster neighborhood sense of identity and responsibility

- 有条件的社区可以引导物业逐步拓展范围，对一定范围的街区进行整体管理，引导成立街区议事会，促进物业、居民、街区单位共管共谋街区发展。
- 定期开展街区公共活动。
- 提供多类型的公益服务。

- The communities with ready conditions can guide the property management to gradually expand the scope, carry on the overall management for a certain range of the community, guide the establishment of community council, and promote the property management, residents and entities in the community to jointly plan the development of community.
- Carry out public activities of the community periodically.
- Provide public welfare services in various forms.

图2-9-14 2019年8月昆山中华园宜居街区议事会成立，为多方协商沟通搭建了基层平台 / In August 2019, Kunshan Zhonghua Garden Community Council was established, as a primary level platform for multi-party consultation and communication | 图片来源：编写组自摄

图2-9-15 居民的需求可以更直观更准确地传达到社区、街道，有效调动了居民参与街区建设的积极性 / The demands of residents can be more intuitively and accurately conveyed to communities and subdistrict, effectively activating the enthusiasm of residents to participate in community construction | 图片来源：编写组自摄

图2-9-16 南京阅江楼省级宜居示范街区共建联盟大会聘请大学教授、社区居民作为“社区规划师” / Co-creation alliance conference of Yuejiang Tower provincial level demonstration community in Nanjing University professors and community residents are invited as “community planning engineers” | 图片来源：编写组自摄

图2-9-17 专家学者组成的“阅江楼宜居街区创建工作坊”形成“共驻、共建、共治、共享”良好格局 / The “Yuejiang Tower Community Creation Workshop” formed by experts and scholars form a good pattern of “joint residing, construction, governance and sharing” | 图片来源：编写组自摄

9.6 划零为整，统一物业管理
Unify diversified property management

有条件的地区可进一步推动老旧小区的成片改造
Areas with ready conditions can further promote the renovation of old communities in batches

· 通过空间整合，完善街巷微循环，满足居民出行、街区消防等要求。

· Complete the micro-circulation of streets and lanes by integrating space, to meet the demands of residents traffic and community fire protection.

方式1：退让用地边界，增加街巷。
疏通街巷，完善微循环交通。
拓宽街巷，满足消防需求。

Method 1: Move back from land boundaries for more streets and lanes.
Get through streets and lanes to complete traffic in micro-circulation.
Widen streets and lanes to meet fire protection requirements.

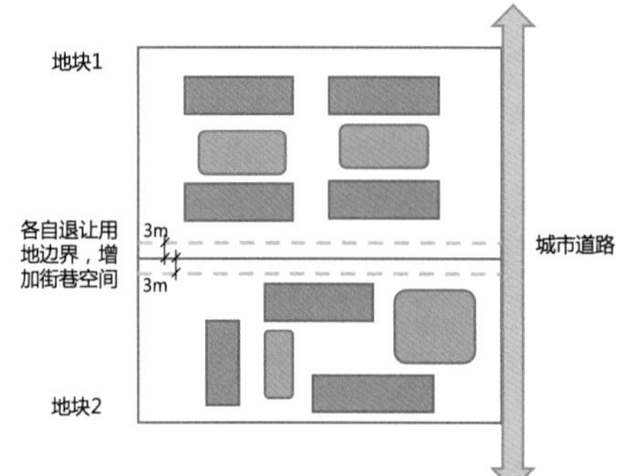

图2-9-18 退让用地边界，增加街巷示意图 / Give way to the boundary of land use and add the schematic diagram of streets and lanes | 图片来源：编写组自绘

图2-9-19 拓宽街巷，满足消防通道要求 / Widen streets and lanes to meet firefighting passage requirements | 图片来源：编写组自摄

方式2：打通断头巷，增加连通度。
连接街巷通道，由线成网，衔接城市道路。
研究制定补偿机制，对各权属单位减少的土地量进行相应补偿。

Method 2: Get through lane with dead end, for more connectivity.
Connect street and lane passages, to form a network, and connect with urban road.
Study and work out compensation mechanism and make corresponding compensation for the amount of land reduced of each ownership entity.

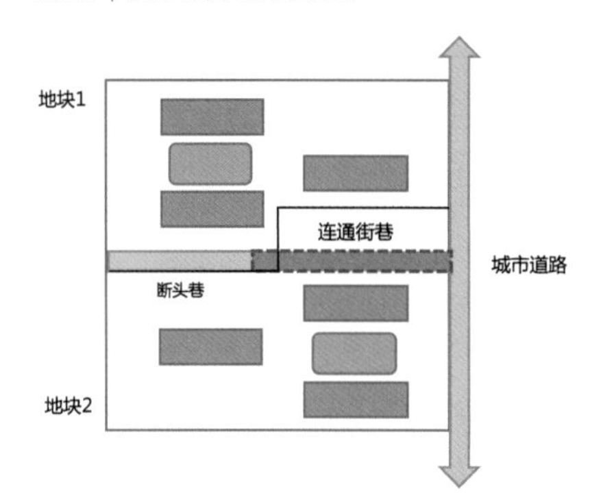

图2-9-20 打通断头巷，增加连通度示意图 / Schematic diagram of breaking through the broken end lane and increasing the connectivity | 图片来源：编写组自绘

图2-9-21 完善街巷微循环以及消防疏散系统 / Complete street and lane micro-circulation and the fire protection evacuation system | 图片来源：编写组自摄

有条件的地区可进一步推动老旧小区的成片改造

Areas with ready conditions can further promote the renovation of old communities in batches

- 适当合并零散小区出入口，统一安防管理。
- Entrances and exits of scattered quarters can be combined as appropriate for unified security management.

零散小区出入口过多，不便管理，且影响城市交通。

Too many entrances and exits of scattered quarters are not conducive to management and affect traffic in the city.

整合零星分散的小区（独栋住宅楼），形成适宜规模的小区空间，统一门卫管理。

Scattered living quarters (separate residence buildings) are integrated into community of appropriate sizes, for management with unified gate security guards.

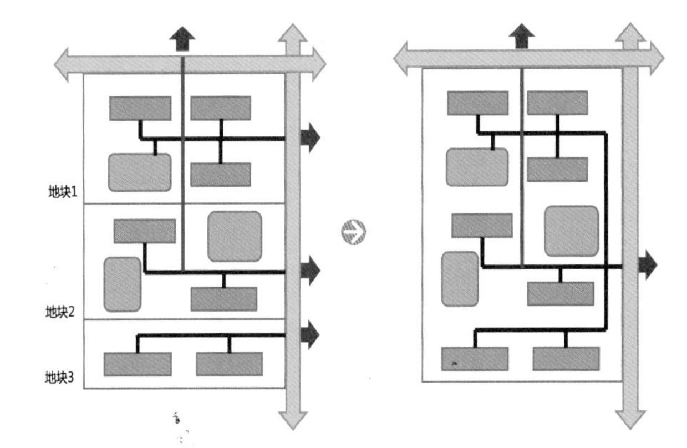

图2-9-22　零散小区空间整合示意图 / Schematic diagram of integrating scattered living quarters | 图片来源：编写组自绘

东台东坝新村小区通过化零为整，整合出入口，将沿街侧门由4个减少为2个，在此基础上完善门禁系统。

Dongba New Village of Dongtai integrated the entrances and exits, reduced the number of side gates along street from 4 to 2, and improved the access control system on this basis.

图2-9-23　东台东坝新村小区空间整合图 / Spatial integration map of Dongba new village community in Dongtai | 图片来源：编写组自摄

Construction Guidance for

Renovation of Old Communities in Jiangsu

江苏老旧小区改造建设导引

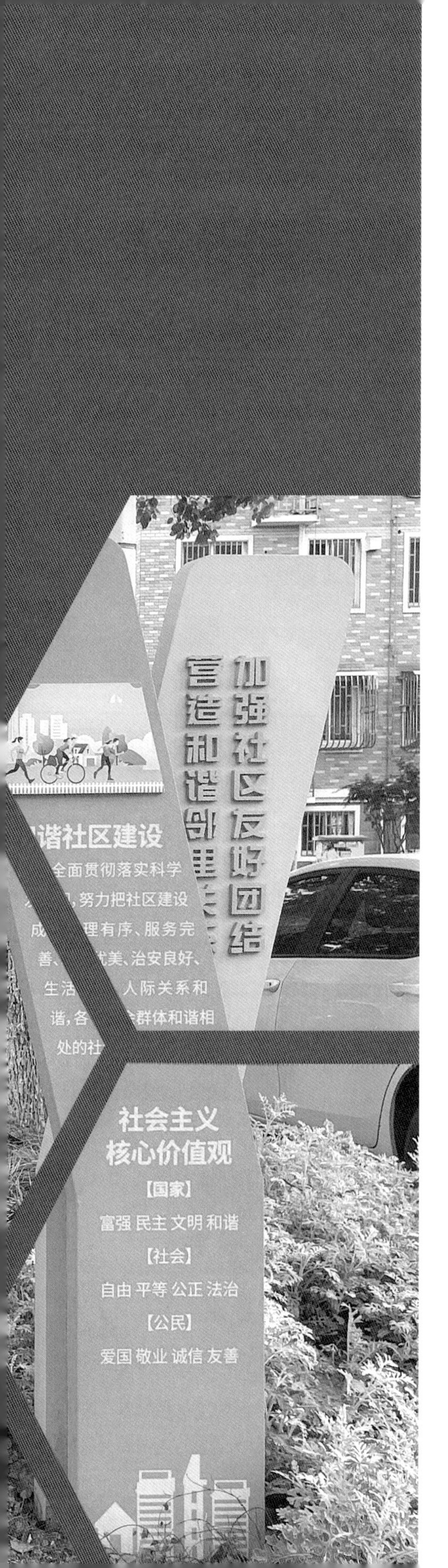

10

建立长效机制

Establish Long-term Mechanism

- 10.1 提供党群服务（社区服务）中心
  Set up Party and mass service (community service) center
- 10.2 提供多样化的信息沟通渠道
  Provide diversified communication channels
- 10.3 培育志愿者队伍
  Foster teams of volunteers
- 10.4 引入社区设计师
  Introduce community designers
- 10.5 鼓励引入专业社区运营机构
  Encourage the introduction of professional community operation institutions
- 10.6 广泛运动社会力量
  Extensively mobilize social forces

10.1 提供党群服务（社区服务）中心
Set up party and mass service (community service) center

提供党建工作、活动的场所
Provide venues for Party building work and activities

- 原则上以社区为单位，设立一处党群服务（社区服务）中心，建筑面积应满足社区党群服务、党群活动需要。
- 鼓励结合社区用房或利用小区存量空间设置，完善设施配置，并体现一定的社区文化特色。

- In principle, a Party and mass service (community service) center should be set up for each community, with floor space meeting the needs of community Party and mass service and activities.
- It is encouraged to complete facilities with community premises or using inventory space in the community, and also to reflect features of community culture.

图2-10-1　昆山枫景苑利用沿街建筑底层空间改造党群服务中心 / The ground floor space of building along street is transformed as a Party and mass service center in Fengjing Garden, Kunshan | 图片来源：枫景苑A区老旧小区改造项目成果

图2-10-2　昆山基层党建带领群众参与各类活动 / Primary level Party organizations in Kunshan lead masses in various activities | 图片来源：亭林城市管理办事处辖区老旧小区改造工程项目活动

发挥基层党建的引领作用

Give play to the leading role of primary level Party building

- 强化党建工作的带动效应，党员干部、老党员等应当发挥示范、动员、协调的作用，结合网格化管理、虚拟党委、党建网络等信息平台，创新党建服务阵地，统筹做好各项群众工作，推动社区治理能力的提升。
- 党建工作应加强与物业的联合，与物业以“一家人”的身份共同开展社区治理，为居民提供细致化服务，将社区服务做细、做深、做实，组织丰富多彩的党群活动，推动社区治理精细化。

- Enhance the driving effect of Party building, Party cadres and veteran Party members should play the role of demonstration, mobilization and coordination, innovate the service position of Party building via information platforms such as grid management, virtual Party committee and Party building network, make overall plans for all kinds of mass work, and promote the upgrading of community governance ability.
- Union of Party building work with property management should be strengthened, to jointly carry out community governance as “one family”, to provide refined service for residents in a practical manner, organize colorful Party and mass activities and further refine the community governance.

图2-10-3 省规划设计集团技术中心党支部与姚坊门物业公司党支部结对共建 / The Party branch of Jiangsu provincial planning and design group and the party branch of yaofangmen property company were built in pairs | 图片来源：编写组自摄

图2-10-4 党支部共建活动之一：宜居街区环境整治 / One of the co-construction activities of the party branch environmental improvement of livable neighborhoods | 图片来源：编写组自摄

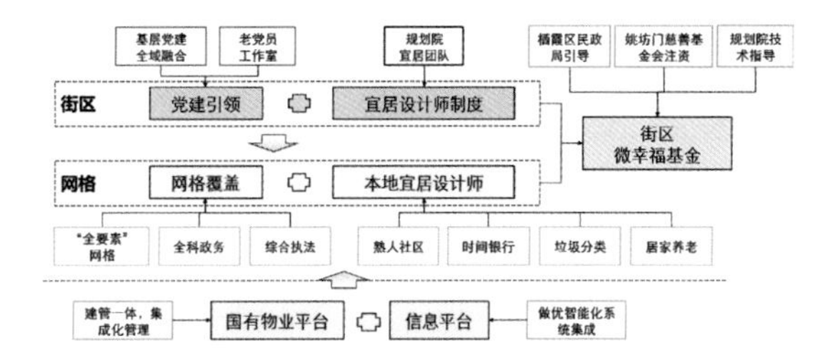

图2-10-5 南京姚坊门宜居街区以党建为引领的长效治理模式 / A long-term governance model led by party building in Yaofangmen livable neighborhood, Nanjing | 图片来源：编写组自绘

10.2 提供多样化的信息沟通渠道
Provide diversified communication channels

保障信息沟通畅通
Ensure unimpeded information communication

- 信息沟通渠道应充分考虑建设成本、居民接受方式等因素，合理提供。
- 社区居民能够畅通接收社区公开信息，包括宣传栏、电子显示屏等传统发布方式。
- 社区居民能够有效反馈社区治理意见，包括议事会、意愿墙、对社区服务打分评价等方式。
- 鼓励社区使用微信、QQ、钉钉等低成本平台软件，也可以开发App或其他手段，创新沟通渠道。

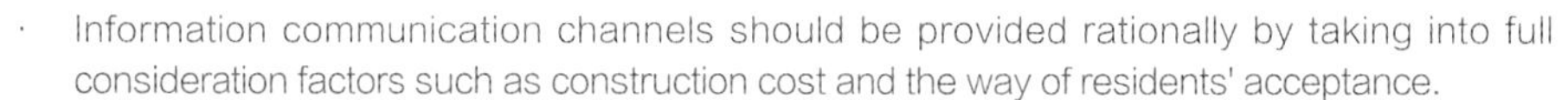

- Information communication channels should be provided rationally by taking into full consideration factors such as construction cost and the way of residents' acceptance.
- Community residents can receive community public information unimpeded, including bulletin boards, electronic display screens and other traditional release methods.
- Community residents can effectively feedback the opinions on community governance, including the discussion panel, wall of wish, rating and evaluation of community services, etc.
- Communities are encouraged to use WeChat, QQ, DingTalk and other low-cost platform software, or develop App or other means, to innovate communication channels.

图2-10-6 南京宁海路街道天津新村社区信息布告栏 / Information board in Tianjin New Village of Ninghai Road Subdistrict, Nanjing | 图片来源：编写组自摄

应有专人维护信息沟通渠道
Here shall be designated personnel to maintain the information communication channels

- 社区应做到及时发布消息、交流沟通、反馈问题。
- 线上无法有效沟通时，应当面对面沟通，保持微笑与耐心。

- A community should ensure timely release of messages, communicate and give feedback.
- When effective communication cannot be made online, communication should be made face to face, with smile and patience.

南京市栖霞区“掌上云社区”：
“掌上云社区”是南京市栖霞区依托微信群、微信公众号，由社区党组织领导，协同居委会、社区居民、物业、社会组织等多元主体围绕服务大多数群众搭建的城市协同共治智慧平台。
“掌上云社区”具有八大功能系统：1.党建引领；2.信息交流；3.“不见面”服务；4.智能回复；5.工单流转；6.多群管理；7.协商议事；8.大数据分析。

“Cloud Community on Palm” of Qixia District, Nanjing:
“Cloud Community on Palm” is an urban collaborative governance smart platform built by Qixia District, Nanjing relying on WeChat group and WeChat official account, led by the community Party organization in collaboration with the neighborhood committee, community residents, property management, social organizations and other diverse subjects for the service of the majority of the people.
“Cloud Community on Palm” has eight functional systems: 1. Party building guidance; 2. Information exchange; 3. online service; 4. smart reply; 5. work sheet circulation; 6. multi-group management; 7. consultation and discussion; 8. big data analysis.

图2-10-7 “掌上云社区”使用主体 / Main users of “Cloud Community on Palm” | 图片来源：栖霞区民政局“掌上云社区”简介

图2-10-8 “掌上云社区”八大功能 / Eight functions of “Cloud Community on Palm” | 图片来源：栖霞区民政局“掌上云社区”简介

图2-10-9 “掌上云社区”微信交流群 / WeChat exchange group of “Cloud Community on Palm” | 图片来源：栖霞区民政局“掌上云社区”简介

10.3 培育志愿者队伍
Foster teams of volunteers

社区应积极发展志愿者队伍
Communities shall actively develop teams of volunteers

- 社区结合日常工作，积极发掘小区里的热心人士、兴趣团体，推动成立志愿者服务队伍，为社区（不限于）提供志愿服务。
- 社区应有专人维护志愿者队伍，对接求助者，招募志愿者。
- In routine work, communities shall actively explore community enthusiasts and interest groups, promote the establishment of volunteer service team, to provide voluntary services for the community (and others).
- A community shall have specially-assigned personnel to maintain the volunteer team, to contact with those in need, and recruit volunteers.

策划志愿者服务日定期提供服务
Plan volunteer service day to provided periodical services

- 服务项目应以社区居民需求为导向，定期开展诸如治安、保健、生活服务、小区微改造、环境维护、社区文化活动等服务。
- Service items should be oriented by the needs of community residents, to carry out regular services such as public security, health care, life services, micro-renovation of community, environmental maintenance, community cultural activities and so on.

图2-10-10 青年志愿者线下物资代购 / Volunteer Service Team purchasing supplies offline | 图片来源：编写组自摄

姚坊门青年志愿服务队：2020年初，突如其来的新冠肺炎疫情席卷全国，全民投入疫情防控。为了给辖区居民提供生活物质保障，金尧花园社区成立姚坊门青年志愿服务队，承担了线上防疫宣传、线下物资代购并送上家门的服务重任。先后帮助居民采购蔬菜、药品等生活物资200余次。防疫形势好转后，青年志愿服务团队继续参与社区各项服务活动，上门送书、送学习用品，关爱青少年健康成长；参与垃圾分类、护绿先行活动，为保护环境出一份力；宣传惠民政策、开展新时代文明实践多彩活动，将志愿服务的理念和温暖传遍花园大家庭的每个角落。

Yaofangmen Youth Volunteer Service Group: At the beginning of 2020, the sudden COVID-19 pandemic swept through the whole country that all the people got involved in the pandemic prevention and control. To provide household supplies to residents under its administration, Jinyao Garden community established the Yaofangmen Youth Volunteer Service Group, which undertook the important services and duties for online pandemic prevention promotion, offline purchases of household supplies and home deliveries. This group have helped residents to purchase vegetables, medicines, and other household substances for over 200 times one after another. After the pandemic situation turned better, the Youth Volunteer Service Group continues to be involved in various community service activities, such as delivering books and school supplies door-to-door and caring the healthy development of teenagers; participating in garbage sorting and pioneering the Green First activities for environment protection; promoting social benefit policies, organizing diverse activities for new era civilization practices, and spreading the idea and warmth of volunteer services to every corner of the big Garden family.

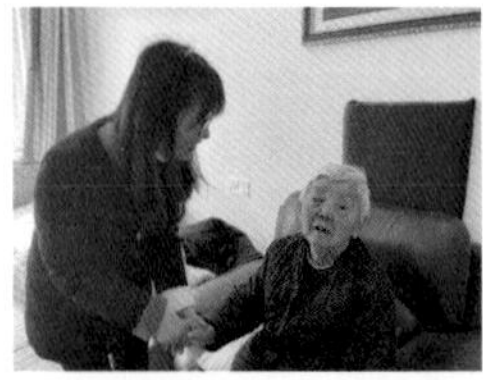

图2-10-11 青年志愿者组织系列活动 / Volunteer Service Team organizing community activities | 图片来源：编写组自摄

10.4 引入社区设计师
Introduce community designers

聘请专业的设计师（团队）共谋社区发展
Invite professional designers (team) to jointly plan community development

- 街道社区结合自身实际，聘请专业设计力量，为社区发展建设提供咨询、设计服务。
- 社区设计师（团队）应在规划、设计、建设等方面具有专业能力，同时具备良好的沟通、组织、协调能力。

- Subdistrict and community shall, according to their own actual need, invite professional design forces to provide consultation and design services for community development and construction.
- Community designers (teams) shall have professional abilities in planning, design and construction, and also be good at communication, organization and coordination.

社区设计师的职责
Duties of community designers

- 参与项目立项、规划、设计、实施的方案审查等。
- 愿意扎根基层，有较强的业务能力、较强的社会责任感，了解城市历史文化，热心公益服务。
- 推动共同缔造行动，培育居民自治能力。

- Participate in project approval, planning, design, and review of implementation of the program, etc.
- Willing to take root at the primary level, have strong professional ability, strong sense of social responsibility, understand the city's history and culture, and be enthusiastic in public service.
- Promote co-creation action, and foster self-governance capability of residents.

成都成华区社区规划师：
成都成华区社区规划师工作制度形成了三级社区规划师队伍，包括导师团、规划设计师和众创组。
导师团负责编制相关技术指南和设计成果评审，并对社区居民等进行规划设计的实操培训。
规划设计师为社区发展提供规划设计专业指导和服务，牵头编制所在社区品质提升项目设计方案等。
Community planners of Chenghua District, Chengdu:
The working system of community planner in Chenghua District of Chengdu has formed a three-level community planner teams, including the mentor group, planning designers and the mass creation group.
The mentor group prepares relevant technical guidelines and review design deliverables, and conduct practical training on planning and design for community residents.
The planning designer provides professional guidance and services for community development, and takes the lead in compiling design schemes for quality improvement projects in the community.

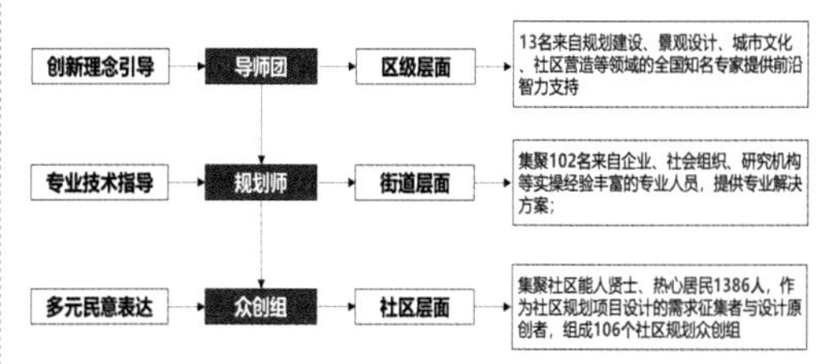

图2-10-12 成都成华社区三级社区规划师队伍 / Third-level community planner team of Chenghua District, Chengdu | 图片来源：根据《城市社区更新理论与实践（成都卷）》编写组自绘

10.5 鼓励引入专业社区运营机构
Encourage the introduction of professional community operation institutions

▶ 有条件的小区可引入专业社区运营机构
Communities with ready conditions can invite professional community operation institutions

· 结合小区的长远发展，有条件的小区可以引入专业社区运营机构。
· 专业机构具有盘活资本、提供服务、减少社区基层压力等优势。
· Communities with ready conditions can invite professional community operation institutions for the long-term development of the community.
· Professional institutions have the advantages of invigorating capital, providing services and reducing pressure at the community level.

图2-10-13 恩派针对疫情推出的疫情危机自救建议 / NPI's self-assistance advice on the epidemic crisis | 图片来源：恩派公益提供

▶ 因地制宜提供专业服务
Provided professional services according to local conditions

· 专业社区运营机构应当以社区、居民的需求为导向，因地制宜地为社区提供服务。
· 提供服务内容涵盖养老、教育、环保、青少年发展、扶贫、助残、社区服务、社会工作等诸多领域。
· Professional community operation institutions shall, oriented by the demands of communities and residents, provide services to communities according to local conditions.
· The services provided shall cover various areas such as old-age care, education, environmental protection, adolescent development, poverty alleviation, assistance to the disabled, community services, and social work.

社区专业运营机构恩派（NPI）：
以恩派公益为例，致力于公益组织孵化、公益人才培养、社区营造与社区服务规模化、公益资金管理、社会企业投资、公共空间运营等业务领域。
A community professional operating agency-NPI:
Taking NPI Charity as an example, it is committed to the incubation of non-profit organizations, the cultivation of public welfare talents, the scaling up of community revitalization and community service, the management of public welfare fund, the investment of social enterprises, and the operation of public space etc.

图2-10-14 恩派公益承接党建项目 / NPI undertakes a public benefit Party building project | 图片来源：恩派公益提供

图2-10-15 恩派吸引汇丰银行注资 / NPI attracts investment from HSBC | 图片来源：恩派公益提供

图2-10-16 恩派策划实施公益展会 / A public welfare exhibition planned and implemented by NPI | 图片来源：恩派公益提供

图2-10-17 联合举办大学生社会创新挑战赛 / Jointly sponsor social innovation challenge competition of university students | 图片来源：恩派公益提供

图2-10-18 恩派托管市民中心 / A citizen center managed by NPI | 图片来源：恩派公益提供

图2-10-19 恩派自主研发的公共空间运营工具 / The public space operation tool developed by NPI | 图片来源：恩派公益提供

10.6 广泛动员社会力量
Extensively mobilize social forces

广泛调动社会力量
Extensively mobilize social forces

- 除了政府和企业外，广泛调动社会资源，包括高校、研究机构、社会团体、志愿者、基金会等部门。
- 社会可以贡献智力、资金、人力等多方面力量。
- In addition to governments and enterprises, social resources shall be extensively mobilized, including universities, research institutions, social organizations, volunteers, and foundations.
- The society can contribute intelligence, capital, manpower and other forces.

图2-10-20 社会团体组织居民确定项目 / Social group organizes residents to determine a project | 图片来源：李郇《社区治理与社区规划的实践——共同缔造工作坊》

创新社会资源参与机制
Innovate social resources participation mechanism

- 鼓励各地创新社会资源参与机制，例如自下而上议题形成机制、以奖代补激励机制、社会组织备案制度、社会组织孵化培育制度等。
- 社会资源参与可以更好地形成良好的社会氛围。
- It is encouraged to innovate social resource participation mechanism, such as the mechanism for forming subjects from bottom to top, the incentive mechanism with awards instead of subsidies, the filing system for social organizations, and the incubation and cultivation system for social organizations.
- Participation of social resources can help form a good social atmosphere.

图2-10-21 大学生参与共同缔造工作坊 / University students participate in co-creation workshops | 图片来源：李郇《社区治理与社区规划的实践——共同缔造工作坊》

多巴安教育：
多巴安是国内公益组织能力建设与人才发展数字化解决方案供应商；多巴安运用大数据、人工智能等技术，帮助公益组织建立自己的在线训练与员工赋能平台；多巴安是恩派旗下的人才发展与智能学习SaaS的服务系统；多巴安将智能、社交以及游戏应用于公益组织人才发展平台的搭建，为公益组织成长赋能。
DUOBAAN Education:
DUOBAAN is a digital solution provider for the capacity building and talent development of domestic non-profit organizations; DUOBAAN uses big data, artificial intelligence and other technologies to help non-profit organizations build their own online training and employee empowerment platform; DUOBAAN is a talent development and intelligent learning SaaS service system under NPI; DUOBAAN applies intelligence, social interaction and games to the establishment of the talent development platform for non-profit organizations, to further empower their growth.

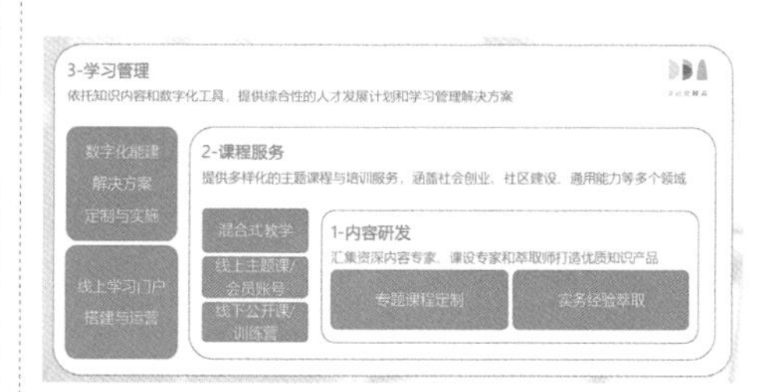

图2-10-22 多巴安教育提供不同层次的能力服务 | 图片来源：多巴安教育信息科技(北京)有限公司提供

图2-10-23 LOGO和公众号二维码 / LOGO and QR code for official account | 图片来源：多巴安教育信息科技(北京)有限公司提供

PART 03 | 工作指引

Work Guidelines

1 科学规划 Scientific planning

· 城镇老旧小区是城市更新中至关重要的社会与空间“双重提升”的机遇地带，其更新改造不能局限于一个个独立的工程性项目就事论事，而需要站在城市统筹发展的高度，科学规划，发挥其促进城市社会空间结构优化和土地价值提升的作用。

· The old communities in cities and towns are an important opportunity zone for “double upgrading” of society and space in urban renewal. Its renewal and renovation can not be limited to independent individual engineering projects, but make scientific planning from the perspective of urban overall development, and play its role in promoting the optimization of urban social spatial structure and the promotion of land value.

· **摸清底数：**通过深入细致的现场勘察和居民问卷调查，从现存的住宅质量状况、基础设施供应、公共设施配套、开敞空间利用等物质环境，以及物业管理、业主组织、居民对更新的态度等社会人文环境进行全面摸底调查和问题评估诊断，摸清既有城镇老旧小区底数，建立现状调查基础数据库。

· **Conduct survey:** through in-depth and detailed field investigation and questionnaire survey of residents, a comprehensive investigation and problem assessment and diagnosis are conducted from the perspective of physical environment such as existing residential quality, infrastructure supply, public facilities, and open space utilization etc, as well as the social and cultural environment such as property management, owners' organizations and residents' attitude towards renewal, so as to find out the truth of the existing urban old residential areas, and establish the basic database of status quo survey.

· **制定计划：** 根据现状摸底评估，综合分析研判老旧小区的物质和社会环境状况，科学编制改造规划；结合居民改造意愿、社会参与度、财政能力等因素，区分轻重缓急，分批建立实施项目库，制定年度计划，优先将居民改造意愿强、参与积极性高的纳入年度计划；充分做好前期工作，推动老旧小区改造有序开展。

· Make plans: according to the status quo survey, make a thorough assessment, comprehensively analyze and judge the material and social environment conditions of the old community, and prepare the scientific renovation plan. According to the residents' willingness to innovation, social participation, financial ability and other factors, give different priorities, set up and implement project database in batches, and make annual plans; priority should be given to those who have strong willingness for renovation and high participation enthusiasm. Carry out the preliminary work, and promote the orderly development of the renovation of old communities.

· **分类施策：** 各地应根据自身实际情况，充分考虑老旧小区在城市空间结构和原有社会网络中的基础，结合老旧小区的衰败状况、人口特点、改造诉求等因素，因地制宜制定差异化的改造策略，运用多种途径和多种手段对城镇老旧小区进行综合治理和更新改造。

· Implement policy in a classified type: all localities should, according to their own actual situation, take full account of the basis of the old community in the urban spatial structure and the original social network, as well as the decline of the old community, population characteristics, and renovation demands and other factors, formulate differentiated renovation strategies according to local conditions, and adopt a variety of ways and means to comprehensively manage and renovate the old communities.

2 工作流程 Work flow

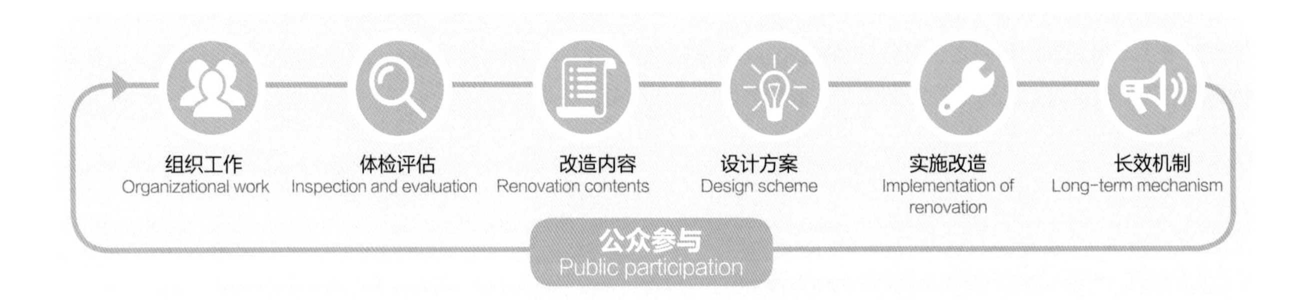

组织工作 Organizational work

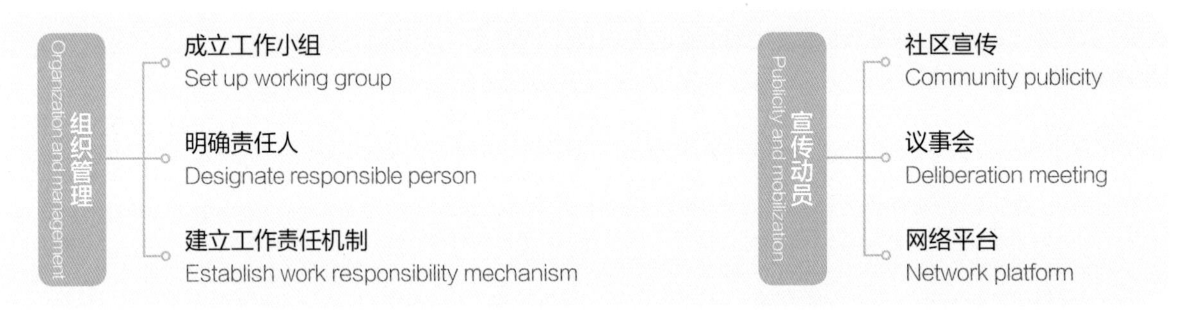

体检评估 Inspection and evaluation

改造内容
Renovation contents

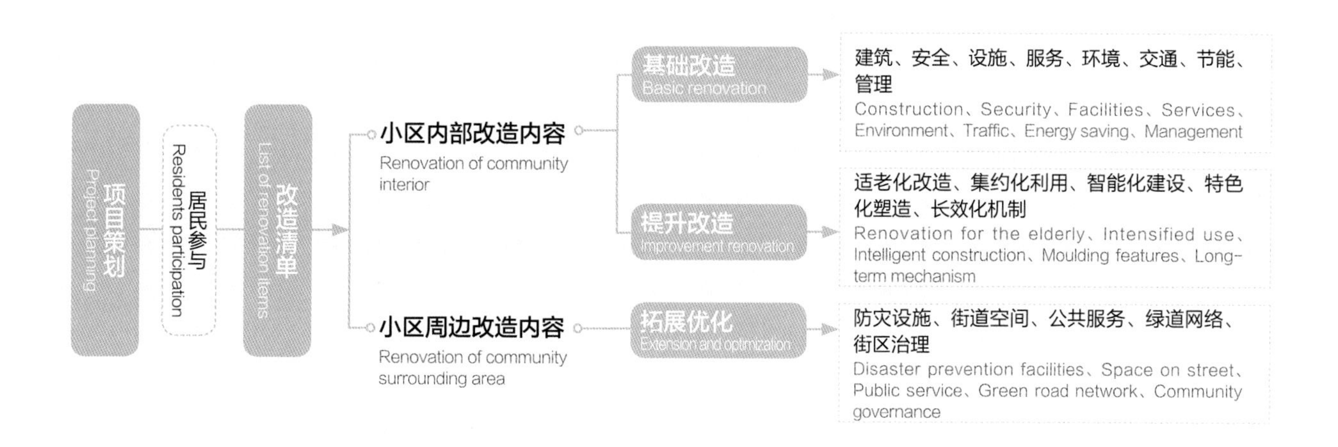

设计方案
Design scheme

实施改造
Implement renovation

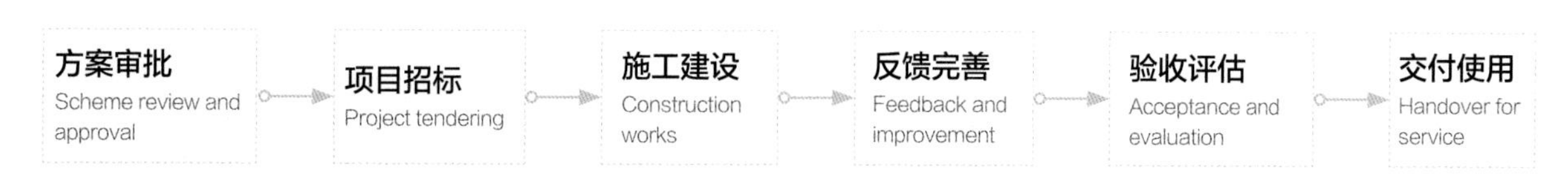

长效机制
Long-term mechanism

3 体检评估
Inspection and evaluation

调研方法
Investigation method

调研方法

调研方法	针对人群 / 空间	调研目的	调研手段
主动采访法	小区各类人群	了解不同人群对小区的看法以及改善诉求	问卷调查 部门走访 人群访谈
实地观察法	建筑、设施 交通、公共空间	了解建筑质量、违章搭建、道路交通、停车及公共设施配置及使用状况、公共空间的使用人群、活动内容、空间活力等内容	实地踏勘 PSPL 分析 分时观察
活动调研法	小区各类人群 特色资源	侧面了解小区及周边街区的特色资源（包括物质与非物质）、特定人群的需求、对街区改善行动的意见	策划各类主题活动
大数据分析法	小区人群活动规律	客观了解小区及周边街区人群的活动规律与功能需求	热力分布 词频搜索 APP 数据采集

Investigation method

Investigation method	Target people/ space	Investigation purpose	Investigation means
Active interview	Various people in the community	Know the views and improvement appeals of different people for the community	Questionnaire Visit departments Interview with people
Spot observation	Buildings and facilities，Traffic and public space	Know the building quality, illegal setups, road traffic, deployment and use of parking and public facilities, people using the public space, activity contents and space vitality	Spot survey PSPL analysis Observation at different time
Activity investigation	Various people in the community，Featured resources	Know indirectly the features resources (both tangible and intangible) of the community and its surrounding area, demands of specific people and comments on improving actions of the community	Plan various theme activities
Big data analysis	Activity regularity of people in the community	Know objectively the activity regularity and function demands of people in the community and surrounding areas	Heat distribution Word frequency search APP data acquisition

宜兴东氿新城宜居街区建设行动——面对面的访谈
Action for building livable community for Dongjiu New Town in Yixing: Face-to-face interview

- 根据改造目标精心策划访谈内容，分别面向住宅小区居民、使用公共空间的人群、使用公共设施的人群、商业区的人群进行访谈和发放问卷，以深入了解人们的问题和诉求。
- The interview contents were carefully planned according to the renovation objectives, and interview was conducted with and questionnaires distributed to residents in the community, people using the public space and using public facilities and people in the commercial areas, to know deeply questions and appeals of the people.

关注小区各类人群的不同需求

图3-0-1 针对业主的访谈 / Interview with owners | 图片来源：编写组自摄

图3-0-2 针对儿童的访谈 / Interview with children | 图片来源：编写组自摄

图3-0-3 针对老人的访谈 / Interview with the elderly | 图片来源：编写组自摄

3 体检评估
Inspection and evaluation

小区内部体检
Internal inspection in the community

小区内部体检

体检要素	主要内容	评估分析
房屋	建筑结构，屋顶、外墙，以及楼道门窗、楼梯踏步、扶手等公共部位	有无危房、有无违章搭建，建筑外墙有无渗漏破损，楼道内部有无安全隐患
基础设施	供水、排水、供电、供气、照明设施	设施有无缺项和安全隐患，是否实施雨污分流和管线下地，设施能否满足居民生活需求
消防	消防设施、电气火灾报警装置、消防通道	消防设施设置是否满足相关标准规范，设施功能是否完好，消防通道是否畅通
安防	小区及入户门禁、安防监控设施、安全疏散标识、避难场所	小区有无安保人员管理，门禁系统是否功能完好，安全标识是否清晰、齐全
社区服务	居家养老服务、日间照料、托育服务	设施规模和布点是否满足相关规范要求
便民设施	快递柜、室外晾晒设施、休息座椅、住宅楼牌、门牌、道路指示牌、公厕等	设施有无缺项，是否功能完好、标识清晰，可达性是否良好
无障碍设施	小区和单元门出入口、小区活动场地出入口无障碍设施	无障碍设施的功能和布点是否满足需求
活动场地	小区内部集中活动场地以及健身器材、儿童游乐设施等	场地面积是否足够，设施有无缺项和安全隐患，能否满足活动需求
环卫	公共空间卫生、垃圾分类情况、垃圾回收点设置等	生活垃圾能否日产日清，垃圾分类设施是否住区全覆盖，楼道等公共空间是否干净整洁
绿化	小区公共绿地、行道树，市政设施围护	有无集中公共绿地，小区绿地率是否满足要求，植物是否得到妥善养护和管理，市政设施是否采取隐蔽化、景观化等举措

◎ 以房屋体检为例
An example of house inspection

· 通过现场踏勘、访谈等方式进行充分调查，逐一拍照记录，并进行综合评价，重点关注存在安全问题的建筑。

· Make full investigation in the forms of spot survey and interview, and record with photos one by one, and comprehensive assessment shall be made, with focus on buildings with safety problems.

Internal inspection in the community

Element	Main content	Evaluation and analysis
Houses	Building structure, roof, exterior wall, and public parts such as stairway doors and windows, stair steps and handrails	Dangerous house, illegal setups, leakage or damage of building exterior wall, and safety peril in the stairway
Infrastructure	Water supply and drainage, power and gas supply, and lighting facilities	Any missing item or safety peril, separation of rainwater from sewage water, pipeline buried under ground, and whether the facilities can meet the demand of residents in life
Fire protection	Firefighting facilities, electrical fire alarm devices, and firefighting passages	The conformity of firefighting facilities to relevant standards and specifications, the sound functions of facilities and any impediment of firefighting passages
Security guard	Community and house access control, security monitoring facilities, security evacuation signs and sheltering venues	Management by security personnel for the community, the good functions of access control system, and clear and complete safety signs
Community services	Home-based care for the aged, day care and nursery services	The conformity of sizes and distribution of facilities to the relevant specifications
Convenient facilities for people	Express mail cabinets, outdoor air drying facilities, rest seats, residential building plates, door plates, road signs, public toilets, etc.	Any missing item in facilities, their good functions, clear marking and good accessibility
Accessible facilities	Accessible facilities at the community and unit door entrances and entrances of community activity venues	Whether the functions and distribution of accessible facilities can meet the demand
Activity venues	Activity venues inside the community and body fit apparatus and children recreation facilities	Sufficient area of venues, any missing item or safety peril in facilities, and whether activity demands can be met
Environmental sanitation	Sanitation in public space, waste classification and setup of waste collection points	Whether domestic wastes are removed as they are produced every day, the waste classification facilities can cover the whole community and public spaces such as stairways are clean and tidy
Plantation	Public green land, trees along roads and municipal facilities in the community	Any centralized public green land, conformity of green land ratio in the community to the specification, proper caring and management of plants, and concealing and landscaping for municipal facilities

图3-0-4 无障碍设施不完善 / Imperfect accessible facilities | 图片来源：编写组自摄

图3-0-5 小区环境不佳 / Improper environment inside the community | 图片来源：编写组自摄

图3-0-6 楼道内部杂乱 / Chaos inside stairways | 图片来源：编写组自摄

图3-0-7 底层门面错杂 / Disorder of facade of shops on the ground floor | 图片来源：编写组自摄

3 体检评估
Inspection and evaluation

小区内部体检
Internal inspection in the community

小区内部体检

体检要素	主要内容	评估分析
车行道	小区出入口交通、内部道路系统	出入口交通是否顺畅，是否人车分流，小区路面是否平整
人行道	小区内部步行道及慢行环境	步行通道是否连续和安全，步行环境是否宜人
停车	机动车与非机动车停车场地、停车管理、电动自行车充电桩	是否有规范的停车场地，是否配置电动车充电桩，有无乱停乱放现象
建筑节能	建筑物外墙、门窗保温节能性能、公共照明灯具	节能性能是否完好，是否开展过节能改造
海绵设施	机动车道、人行道、停车场和广场绿地	是否采用透水铺装，雨水是否收集利用
党建宣传	社区党群服务中心或便民服务中心服务情况、信息发布渠道和宣传情况	是否设有服务中心，是否定期发布服务信息并提供服务，居民的满意度
物业管理	物业管理用房配套、管理服务水平、业主委员会运作情况	是否有专业物业管理，是否成立业主委员会，小区日常管理是否满足需求
志愿服务	社区志愿者的组成和服务情况	是否定期开展志愿者服务活动
资金	小区公共资金（财政拨付专款、小区公共收入等）的管理和使用情况	小区公共资金是否得到妥善使用，资金的使用记录是否详实、公开

◎ 以志愿者服务体检为例
Example of inspection of volunteer service

· 针对社区志愿者队伍的培育、发展和服务情况开展评估。

· Evaluation shall be made on the fostering and development and service provided by the community volunteers.

Internal inspection in the community

Element	Main content	Evaluation and analysis
Vehicle lanes	Community access traffic and internal road system	Smooth access traffic, separation of people and vehicle flows, and flat surface of roads in the community
Sidewalks	Sidewalks and slow traffic environment in the community	Whether the sidewalks are continuous and safe, and the walking environment is comfortable
Parking	Parking lots for motorized and non-motorized vehicles, parking management and charge stakes for electric bicycles	Any standardized parking lot, provision of charge stakes for electric bicycles, and any disorderly parking
Building energy saving	The insulation and energy saving performance of building exterior wall and doors and windows, and public lighting fixtures	Good energy saving performance, and any energy saving renovation made
Sponge facilities	Vehicle lanes, sidewalks, parking lots, squares and green land	Any water permeable paving used, whether rainwater is collected for use
Party building and publicity	Services provided by community Party and mass service center or people service center, information release channels and publicity	Setup of service center, whether service information is issued periodically and service provided, and resident satisfaction
Property management	Rooms for property management, management service level, and operation of owners committee	Any professional property management, the setup of owners committee, and whether community routine management meets the demands
Volunteer service	Composition and service provided by community volunteers	Whether volunteer service activities are carried out regularly
Funds	The management and application of public funds of the community (special funds allocated by the finance and public revenue of the community)	Proper use of public funds of the community, and detailed and transparent records of the application of funds

图3-0-8 志愿者文化活动 / Cultural activities of volunteers | 图片来源：宜兴市长新社区提供

图3-0-9 志愿者便民服务 / Service by volunteers | 图片来源：宜兴市长新社区提供

图3-0-10 志愿者送医服务 / Medical service by volunteers | 图片来源：宜兴市下漳社区提供

3 体检评估
Inspection and evaluation

小区周边体检
Inspection on surrounding areas of the community

小区周边体检

体检要素	主要内容	评估分析
街区安全	街道安防监管情况，以及幼儿园、中小学出入口安全，街区消防、疏散通道、避难场所	街头监控点和街面见警率是否满足需要，是否存在安全盲区，中小学出入口高峰时段安全管理是否满足要求
边角地	街巷边角畸零地	有无街巷边角畸零地，空间是否得到有效利用
步行环境	人行道、健身道、休闲步道	人行道是否安全，步行空间是否连续贯通，有无休息座椅等配套设施，有无能够就近使用的健身绿道
街道景观	沿街建筑风貌、店招广告、街道铺装以及街道绿化情况	沿街建筑立面、店招广告是否风貌协调，街道铺装是否存有破损，行道树是否得到良好养护
街道家具	垃圾桶、灯杆、市政设施、智能充电设施、街头雕塑、标识等	街道家具的布点是否合理，功能是否正常
配套服务设施	卫生服务站或社区医院、老年服务用房、幼儿园或托育场所	设施是否满足规范要求，设置标准是否达标，功能是否齐全，可达性是否良好
便民商业网点	超市便利店、菜场、药店、洗衣店、理发店、银行服务网点等	设施是否满足规范要求，是否满足居民日常使用需求、便捷可达
其他便民设施	公共厕所、公交站点或地铁轨道站点、公共自行车服务点	有无相关设施，设置标准是否满足规范和居民日常使用需求，可达性如何
绿地公园	街头绿地、口袋公园等，以及结合绿化建设的篮球、足球、羽毛球等常规运动设施，以及休息座椅等配套设施	公园是否步行可达，相关运动设施、运动器材以及配套服务设施是否满足使用需求，公园植物是否得到养护和管理
历史文化	文保单位、历史建筑、传统风貌建筑和历史环境要素等历史文化资源	有无相关历史文化资源，是否得到妥善保护和利用
特色活动	街头特色文化活动和文艺演出等	有无定期和不定期的特色活动，居民参与度如何，群众是否满意

以街区安全体检为例
An example of security inspection for community

- 针对街区消防设施、避难场所、疏散标识指引等设施布置和使用情况开展评估。
- Evaluation shall be made on the layout and use of facilities such as firefighting facilities, sheltering venues and evacuation signs and guide of the community.

Inspection on surrounding areas of the community

Element	Main content	Evaluation and analysis
Security of community	Security monitoring for the community, access security of kindergartens, primary and middle schools, firefighting and evacuation passages and sheltering venues for the community	Whether the street monitoring points and police visibility meet the demand, any blind zone for security, whether security management for school accesses at peak hours meets the requirements
Edge and corner land	Edge and corner land by streets and lanes	Any edge and corner land by streets and lanes, and effective use of space
Walking environment	Sidewalks, fitness trails and leisure trails	Safety for sidewalks, whether the walking space is continuous, and provided with facilities such as rest chairs, any fitness green trails available nearby
Street landscape	Building style and shop advertisements along street, street pavement and plantation	Whether building facade and shop advertisements along street are harmonic with style, any damage in street pavement, and proper caring of sidewalk trees
Street furniture	Trash bins, light poles, municipal facilities, smart charging facilities, street sculptures, signs, etc.	Whether the layout of street furniture is reasonable, and their functions are normal
Associated service facilities	Health service stations or community hospitals, rooms for elderly service, kindergartens or childcare venues	Whether the facility meets the specification requirements, the setting is up to standard, the functions are complete, and with good accessibility
Convenient shops and outlets	Supermarket and convenience stores, grocery stores, drugstore, laundry, barber shop, bank service outlets, etc.	Whether the facilities meet the specifications, meet the daily use needs of residents, convenient and accessible
Other convenient facilities for people	Public toilets, bus stops or metro stations, and public bike service points	Any relevant facilities, whether they can meet the specifications and daily use needs of residents, and the accessibility
Green land and park	Street green space, pocket park, and other conventional sports facilities such as basketball, football and badminton combined with green land construction, as well as supporting facilities such as rest seats	Whether the park is within walking distance, the relevant sports facilities, sports equipment and associated service facilities meet the use needs, and the plants in the park are taken care and managed
History and culture	Historical and cultural resources such as cultural protection units, historical buildings, buildings with traditional features and elements of historical environment	Any relevant historical and cultural resources, and whether they are properly protected and utilized
Featured activities	Street featured cultural activities and artistic performances	Any featured activities held periodically or non-periodically, the participation rate by residents, and satisfaction of masses

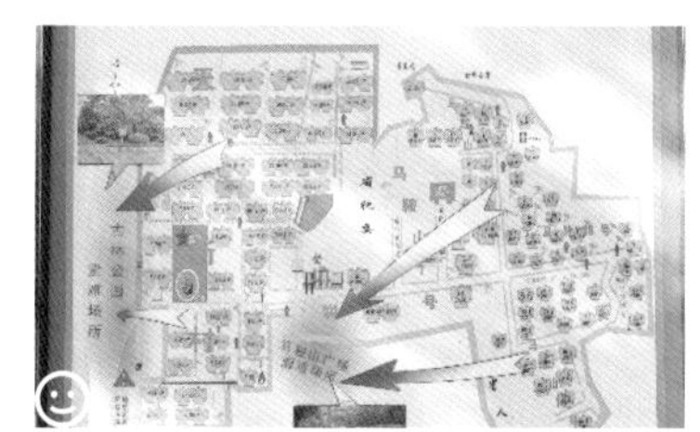

图3-0-11 灾害风险地图 / Disaster risk map | 图片来源：编写组自摄

图3-0-12 规范灾害管理 / Standardize the disaster management | 图片来源：编写组自摄

3 体检评估
Inspection and evaluation

评估标准
Evaluation standard

· 体检评估是老旧小区改造的必要前期工作，通过系统的体检可以找出老旧小区拟待解决的主要问题，明晰未来改造重点。在进行体检评估工作的过程中，应依据标准规范、上位规划、居民评价、人群需求等对各项体检要素进行评价分析。

· Inspection and evaluation is the necessary work before the renovation of old communities, the main problems urgent for solution in old communities can be identified in systematic inspection, to make clear the main points of renovation in the future. In the inspection and evaluation, various elements shall be evaluated and analyzed according to the standards and specifications, up level planning, assessment by residents and demands of people.

标准规范
Standards and specifications

· 主要包含国家和省市级的相关技术标准与规范，是体检评估的主要依据。

· Mainly including the relevant technical standards and specifications of the state and of provincial and municipal levels, as the main basis for inspection and evaluation.

上位规划
Up level planning

· 老旧小区改造是一项综合性强、内容繁复的工作，改造方案应主动与上位规划进行衔接，其中需要衔接的重点包括形态风貌、历史文化保护、城市更新、设施配套等方面。

· Renovation of old communities is a highly comprehensive and complicated task, the renovation scheme shall be well linked with the up level planning, and the main points for links include form and style, protection of historical culture, city renewal and completion of facilities.

居民评价
Assessment by residents

· 居民针对现状住宅、小区环境以及相关设施的布局和使用性能反映的主要问题和诉求，也是评估的重要依据。

· The main problems and appeals from residents on the layout and performance of status quo residents, community environment and relevant facilities are also important basis for evaluation.

人群需求
Demand of people

· 居民对理想居住条件的需求，包括对住房、医疗、教育、文化、体育等各方面日常生活环境以及社区归属感、邻里交往等精神层面的期望也是评估应当考虑的参考要素。

· Residents' demand on ideal living conditions, including their expectation on housing, medical service, education, culture and sports, the daily living environment and the sense of belongings of community and neighbor exchanges are also reference elements to be duly considered in evaluation.

图3-0-13 了解人群需求 / Understand the demand of people | 图片来源：编写组自摄

分析居民评价
Analyze assessment by residents

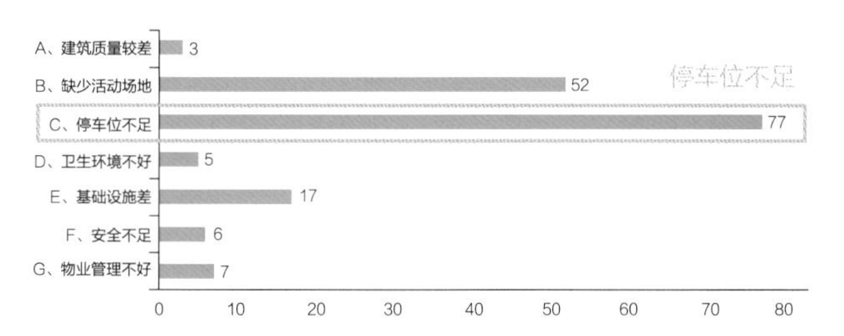

图3-0-14 居民反应的小区现状问题 / Status quo problems of the community as reflected by residents | 图片来源：编写组自绘

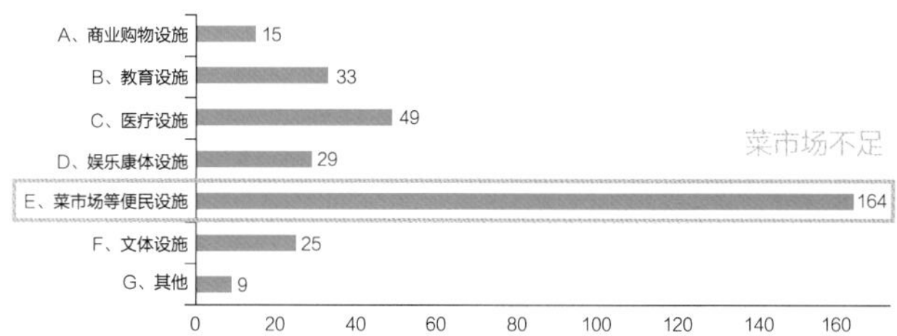

图3-0-15 居民希望增加的服务设施 / Service facilities that the residents hope to be added | 图片来源：编写组自绘

4 共同缔造
Co-creation

共同缔造是什么?
What is co-creation?

· 共同缔造是通过引导、激励、参与、沟通、协调、制度等手段，汇聚社区居民、政府、企业、社会等多方力量，凝聚社区共识、创造社区文化、积极参与社区公共事务的过程，最终形成以居民自治为核心、多方共治为支撑的社区治理体系，以实现环境与社会良性互动、可持续发展的理念与方法。

· Co-creation refers to the process to put together the forces of residents, government, enterprises and the society by means of guiding, stimulation, participation, communication, coordination and rules, to reach common view of the community, create community culture and actively participate in community public affairs, to finally form the community governance system with resident self-governance as the core and supported by multiple parties, to realize the active interaction and sustainable development between environment and the society.

参与主体有哪些?
What are the participating subjects ?

· 参与主体涉及所有利益主体，包括且不仅限于社区居民、街道社区、政府部门、社团组织、设计施工机构、高校、企业、基金会、志愿者等，形成“政府－市场－社会”三位一体的共治主体。

· The participating subjects involve all interest bodies, including and not limited to community residents, communities, government departments, social organizations, design institute and construction companies, universities, enterprises, foundations, volunteers, etc., forming a trinity co-governing subject of the “government-market-society”.

图3-0-16 居民共绘问题地图 / Residents drawing problem map together | 图片来源：编写组自摄

各参与主体在共同缔造中扮演怎样的角色?
What are the roles of participating subjects in co-creation?

- 目前共同缔造最主要的参与主体有四类:
- At present, there are mainly four types of participating subjects in co-creation:

居民从参与到自治
Residents, from participation to self-governance

- 居民从最初的参与到最终实现自治，让社区权力归还居民，需要一个过程。具体共同缔造的方式包括居民议事会、社区能人召集行动、微改造微幸福项目、业主委员会会议等。
- It is a process from the initial participation to final self-governance by residents, to return the power of community to residents. Forms of co-creation include residents deliberation council, community calling actions, micro renovation for happiness projects, and owners committee meeting.

政府从主导到引导
Government, from leading to guiding

- 当前基层政府是改造项目的最重要力量，随着改造工作的深入，政府逐步还权于居民和市场。具体共同缔造方式包括基层走访调研、基层治理改革、出台相关政策、举行赛事活动、示范项目带动等。
- At present, the primary level government is the most important force in the renovation projects. With the deepening of the renovation work, the government gradually returns the power to the residents and the market. Specific ways of co-creation include grassroots visits and investigation, grassroots governance reform, issuing relevant policies, holding competition events, demonstration projects drive, etc.

设计师从设计到咨询
Designers, from design to consultancy

- 设计师作为专业人士可以给出相对合理、完善、专业的指导和建议，并做到全过程咨询。具体方式包括提供专业咨询、具体项目设计、举办工作坊、策划系列活动、组织研讨会、开发数据平台等。
- Designers as professionals can give relatively reasonable, complete, and professional guidance and advice, and do the consultation in the whole process. Specific ways include providing professional consulting, specific project design, holding workshops, planning series of activities, organizing seminars, developing data platform, etc.

社造联盟创新服务
Innovation service by community alliance

- 社造联盟是近年来随着社区营造工作的深入新兴发展的具有市场化特征的组织，具体方式较为多样，在吸引外部资金、孵化社区团体、长效运营管理方面具有优势。
- Community alliance is an organization with marketized feature emerging with the deepening of community creation work in recent years, featuring diversified forms and with advantages in absorbing external funds, incubating community groups and long-term operation management.

4 共同缔造：居民从参与到自治
Co-creation: residents from participation to self-governance

居民为何能自治?
Why residents can do the self-governance?

· 居民是社区的主人，他们每天生活在社区之中享受公共服务、接受公共管理、参与公共事务，因此居民应当是社区公共事务的主导者之一，通过参与社区治理与老旧小区改造，强化主体意识、认识自身责任、学习专业知识、学会沟通交流、发挥群众创意，最终形成社区自治的核心力量。

· Residents are the masters of the community. They live in the community every day to enjoy public services, accept public administration and participate in public affairs, therefore they should be one of the main players in community public affairs. By participating in community governance and renovation of old communities, they can enhance their awareness as main body, know their own responsibilities, learn professional knowledge and communication, give play to the mass creation and finally become the core force of community self-governance.

居民自治需要注意哪些?
What are the precautions in resident self-governance?

· 由于长时间处于缺乏参与的状态，因此从参与到自治是一个漫长的过程，不能急于求成。

· 居民自治的方式、过程需要因地制宜，在推行的过程中培养自治能力，同时需要得到大多数人的支持，避免沦为少数人的游戏。

· As they are not lacking in participation over long time, it is a long process from participation to self-governance, and we must not be anxious for success.

· The form and process of resident governance should be based on local conditions, the self-governance capability should be fostered in the process, and support from the majority is required, it should be avoided to turn it into a game of few people.

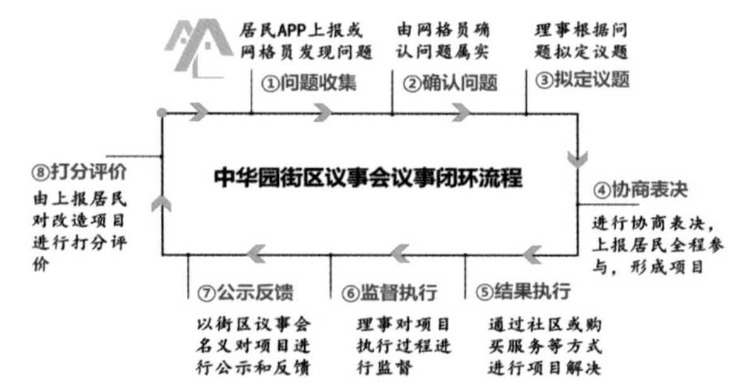

图3-0-17 昆山中华园街区议事会议事流程 / Discussion procedure of deliberation council in Zhonghuayuan Community of Kunshan | 图片来源：编写组自绘

居民如何从参与到自治？
How can residents proceed from participation to self-governance?

- 在目前阶段，虽然社区居民参与意识较弱，但是通过政府、设计师、社造联盟的介入与帮助，平等对话，建立协商合作平台，适量给予奖励，建立小区人才库，设立议事会，逐步激发居民参与热情，扩大居民参与途径。
- 具体路径可从居民最关心的身边事入手，如加装电梯、停车设施增补、楼道美化、小区活动场地改造等，发动居民自己行动，让居民看到自己改造环境的力量，以此逐步推动居民赋能。

- At present, the awareness of participation of residents is weak, but with the intervention and assistance by the government, designers and community alliance, equal dialogue, establishing consultation and cooperation platforms, appropriate incentives, establishing community talent pools and setting up deliberation council, the enthusiasm of residents in participation can be activated, and the ways of participation will be expanded.
- Specifically, it can be started from affairs most concerned by the residents, such as adding elevators, adding parking facilities, stairway beautifying and renovation of community activity venues, residents can be mobilized to act by themselves, so that they see their forces to change the environment by themselves, to gradually enable them in governance.

图3-0-18 居民与设计师共同讨论问题地图 / Residents are discussing on maps with designers | 图片来源：编写组自摄

图3-0-19 就公共领域开展议事活动 / Conduct deliberation activities in public domain | 图片来源：编写组自摄

图3-0-20 选举能人成为议事会班子成员 / Elect able people as members of the deliberation council | 图片来源：编写组自摄

4 共同缔造：居民从参与到自治
Co-creation: residents from participation to self-governance

从微更新着手
Start from micro upgrading

- “微更新”或“微改造”项目是关乎居民利益的身边事，一般改造成本低、改造难度低，却能够显著提高居民的参与度和幸福感。
- “微更新”项目改造资金由政府、居民和其他社会资本共担，政府可以采用“以奖代补”的形式进行鼓励。
- “微更新”项目一般步骤为：居民提出改造需求，结合需求的急迫程度、成本高低等因素进行排序，选出实施项目，居民自己设计、动手进行改造，政府进行奖补。

- “Micro upgrading” or “micro renovation” projects are closely related to the interests of residents, generally at low cost and is not difficult in renovation, but they can substantially raise the participation and sense of happiness of residents.
- Funds for “micro upgrading” renovation are shared by the government, residents and other social capital, and the government can encourage such practices in the form of “bonus as subsidies” .
- Normally a “micro upgrading” project involves such steps: residents proposing renovation demand, selecting projects by ranking with the urgency of demand, cost and other factors, residents doing the renovation by their own design and work, and the government offering bonus as subsidies.

开展系列活动
Carry out series of activities

- 自治组织通过开展系列活动，宣传改造事务，激发居民参与改造的热情，了解群众需求，调解居民难处。
- 活动形式多种多样，传统形式包括居民座谈会、问卷调查、居民入户访谈等；也可以采用包括设计工作坊、设计进校园、关爱弱势群体、结合节庆活动等新兴形式。

- Through a series of activities, self-governance organizations will publicize the renovation affairs, stimulate the residents' enthusiasm to participate in the renovation, understand the needs of the masses, and mediate the difficulties of the residents.
- Activities can be in various forms, including traditional forms of residents discussions, questionnaire survey, residential interviews; new forms can also be adopted, such design workshops, design in schools, caring for vulnerable groups, and those in combination with festival activities.

成立自治组织
Set up self-governance organization

- 由居民或者社区牵头成立自治组织，组织内的成员应当包括多元主体，包括居民、租户、业委会、物业公司、老党员、网格长、社区工作者、经营者、热心公益人士等各类人群代表。
- 自治组织形式多样，可以包括居民议事会、兴趣社团组织、商会组织等多种形式。
- 自治组织应当实现自我管理、自我教育、自我服务，并按照组织的章程、公约进行活动。
- Residents or communities shall take the lead in establishing a self-governance organization, and its members shall include multiple subjects, such as residents, tenants, owners committees, property management companies, veteran Party members, grid leaders, community workers, business operators, public welfare enthusiasts and representative of other groups.
- Self-governance organizations can be in diversified forms, such as resident deliberation council, interests groups and chamber of commerce.
- Self-governance organizations shall realize self-management, self-education and self-service, and hold activities according to the charter and convention of the organizations.

图3-0-21 广场议事会 / Square council | 图片来源：编写组自摄

图3-0-22 组织活动 / Organizing activities | 图片来源：编写组自摄

图3-0-23 红色联席会 / Red joint meeting | 图片来源：编写组自摄

“睦邻”老党员工作室：
尧林仙居社区的“睦邻”老党员工作室成立于2017年9月，“睦”即和好、亲近；“邻”即住处接近的人家又谐音尧林的“林”，寓意邻里友好，又和谐尧林。本着“搭平台，联民心，聚民力”的宗旨，2020年“睦邻”老党员工作室围绕打造宜居社区，优化共治共建开展60余项活动；每月定期开展“红色联席会”，做好居民与社区的纽带工作，传达民情民意，办好民生民事，不断提高居民群众的幸福感、安全感和获得感。
为了更好地倾听群众的所需所求，解决好群众的难点、堵点、痛点问题，社区老党员工作室还开设了“广场议事会”机制，通过面对面的交流沟通，和居民代表现场解决群众提出的问题，解决了碧水苑9幢广场绿化布置和翠林苑长廊建设问题，协调化解了10余次出新疑难矛盾。通过老党员工作室，把矛盾化解在萌芽状态，12345工单从2019年的445条到2020年245条，下降了200条，社区各项工作也渐渐得到了居民的支持和认可。

“Good Neighbor” Veteran Party Member Office:
“Good Neighbor” Veteran Party Member Office of Yaolinxianju Community was established in September 2017. “Good” means harmony and closeness; “Neighbor” means households live nearby and has the same pronunciation with “Lin” in Yaoxin. “Good Neighbor” implies that the neighbors in Yaolinxianju Community maintain harmonious relationship. With the aim of “providing the platform, maintaining close ties with the public, and giving resources of the public into full play”, “Good Neighbor” Veteran Party Member Office in 2020 had carried out more than 60 activities for creating a pleasant community and optimizing co-building and co-governance; held “Red Joint Meeting” every month, maintained close ties between the residents and the community, communicated public feelings and wills, cared about the people’s livelihood, and continuously improved the people’s sense of happiness, security and gain. To better know people’s needs and solve their difficulties and pain points, the Office also created a “square council” mechanism, and had solved the square greening layout of Bishuiyuan #9 Building and the corridor building of Cuilinyuan, and coordinated and settled more than 10 new difficulties and conflicts through face-to-face communication and residents’ representative addressing the problems raised by the residents on site. The conflicts were resolved in the bud through the Office, and the work orders of governmental service hotline 12345 had dropped from 445 in 2019 to 245 in 2020, with a decrease of 200 work orders. The community’s work gradually gets the support and recognition of the residents.

4 共同缔造：政府从主导到引导
Co-creation: government from leading to guiding

图3-0-24 街道和社区召集居民召开议事会 / Street and communities are having meetings with local residents. | 图片来源：编写组自摄

政府是现阶段最重要的参与者
The government is the most important participator in the present phase

- 基层政府作为主导者的身份出现是现阶段社区治理最常见的现象。
- 基层政府尤其是基层党建应当发挥引领和带头作用。
- 政府主导的目的不是掌握更多资源，而是通过更多的资源调配，为居民服务，并培育居民自治。

- Primary level government acting as the leader is the most common phenomenon in community governance at present.
- Primary level governments, especially primary level Party organization should play the leading role.
- The purpose of government taking the lead is to serve the residents and foster resident self-governance through allocation of more resources, instead of mastering more resources.

政府在共同缔造中的角色
Roles of government in co-creation

- 宣传者——解析内涵，激发参与热情
- 触媒者——整合资源，建立多方联系
- 促能者——发现能人，挖掘潜在宝藏
- 发动者——共商共议，发动群众参与
- 组织者——建立组织，凝聚群众力量
- 协助者——协助工作，兼用内外优势
- 引导者——制定制度，引导群众自治

- Propagandist —— analyze the connotation, and stimulate enthusiasm of participation
- Catalyst —— integrate resources and establish multi-party contacts
- Promoter —— find able people and tap potential treasures
- Initiator —— discuss and consult together to mobilize the masses to participate
- Organizer —— Establish organization and put together the forces of the masses
- Assistor —— assist in work to utilize both internal and external advantages
- Guider —— formulate rules and guide the self-governance by the masses

政府需要转变思路，还权赋能
The government needs to change ideas, return the power and enable people

· 政府具有最广泛的资源，因此，引领社区治理具有效率高、动员广、资金多等特点，适合于试点项目的推进。
· 为了追求效率，可能会导致共同缔造流于形式，政府代办过多会抑制居民自治能力。
· 基层政府需要转变思路，通过各种形式激发社会参与热情，如针对社区小微公共空间的赛事活动，吸引市场资源，还权赋能。
· The government has the most extensive resources, therefore it has the features of high efficiency, extensive mobilization and large amount of funds in leading the community governance, suitable for pushing forward pilot projects.
· The pursuit of efficiency may lead to a formalism of co-creation, and too much government agency will inhibit the residents' ability of self-governance.
· Primary level governments need to change ideas, to stimulate social enthusiasm in participation in various forms, such as competitions in community small and micro public space, absorbing market resources, to return power and enable the people.

江苏紫金奖·建筑及环境设计大赛：
· 紫金奖·建筑及环境设计大赛为设计师、学生及社会公众搭建专业性与社会性充分融合的平台，以创意构思和创意设计推动城乡空间品质提升。
· 奖项鼓励真题实做。
· 获得一等奖的作品，有机会与当地政府进行签约，让项目能够落地，推动社区改造，扩大影响力。
Jiangsu Zijin Prize · Grand Design Competition on Architecture and Environment:
· Zijin Prize · Grand Design Competition on Architecture and Environment is a platform built for designers, students and the social public for full merging of profession and sociality, to promote improving the quality of urban and rural spaces with creative conception and designs.
· The prize encourages actual practice of real subjects.
· Works obtaining the first prize has the opportunity to sign a contract with the local government, so that the project can be put into practice, to promote community renovation and expand the influence.

图3-0-25 第六届紫金奖· 建筑及环境设计大赛 / Sixth Zijin Prize · Grand Design Competition on Architecture and Environment | 图片来源：紫金大赛组织提供

图3-0-26 落地签约仪式 / Contract signing ceremony | 图片来源：紫金大赛组织提供

图3-0-27 第六届职业组一等奖作品与参加电视决赛选手 / Works of the first prize in the profession group and players participating in TV final in the sixth session | 图片来源：紫金大赛组织提供

4 共同缔造：设计师从设计到咨询
Co-creation: designer from design to consultancy

图3-0-28 设计师与居民耐心沟通 / Designers communicate pationtly with residents | 图片来源：编写组自摄

设计师是改造过程中的专业力量
Designers are the resources forces in renovation

· 设计师要突破传统的自上而下的设计思路，摆脱“精英式”规划，放低身段，走到基层一线，与基层政府、居民等多方充分沟通，了解需求，掌握问题。

· 设计师的职能不仅仅局限于规划设计，而是要贯穿改造的全流程，提供全过程咨询服务。

· 设计师投身社区规划，也是再教育和再学习的过程，需要提升与社会对话的能力，以及对“以人为本”情怀的坚持。

· Designers shall break the traditional design ideas of proceeding from the top to bottom, get rid of the “elite” planning, go to the front line on the primary level, make full communication with the primary level government and residents, to know the demands and problems.

· Functions of designers are not limited to planning and design, instead, they should provide full-process consultancy service throughout the whole renovation.

· Designers doing the community planning is also a process of re-education and re-study, and this requires high ability on dialogue with the society and the persistence of “putting people first”.

设计师在共同缔造中的角色
Roles of designers in co-creation

图3-0-29 居民座谈会 / Seminar of residents | 图片来源：编写组自摄

· 学习者——广泛学习，增强实践能力
· 组织者——拟定计划，推进规划过程
· 宣传者——图文并茂，解说美好家园
· 沟通者——收集意见，促成多方交流
· 引导者——传授技能，提供专业支持
· 规划者——总结成果，促进学科发展

· Learner —— extensive study to enhance the ability of practice
· Organizer —— work out plans to promote the planning process
· Propagandist —— illustrate the beautiful home with both pictures and words
· Communicator —— collect opinions to promote multi-party exchange
· Guider —— teach skills and provide professional support
· Planner —— sum up results and promote the development of the discipline

南京“宜居姚坊门省级宜居示范街区暑期工作营”：

· 工作营吸引了大学生、小学生、设计师、居民、专家、政府工作人员共同参与。

· 工作营举行了跨界专家讲座、室内作业、户外调研、全民开放日、第三方机构调研、小学生演讲等多种形式的活动。

“Summer Working Camp of Yaofangmen Provincial Level Livable Demonstration Block” in Nanjing:

· The camp attracted college students, primary school students, designers, residents, experts and government working personnel to participate.

· Various activities were held in the camp, including cross-border experts lecture, indoor work, outdoor investigation, Open Day for all, investigation by third-party organizations, and speeches by primary school students.

图3-0-30 招募海报 / Recruiting poster | 图片来源：编写组自摄

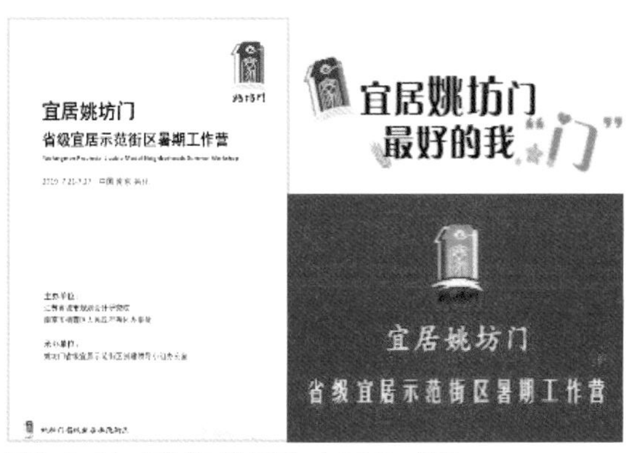

图3-0-31 工作营工作手册、LOGO、旗帜 / Work manual, LOGO and flags of working camp | 图片来源：编写组自摄

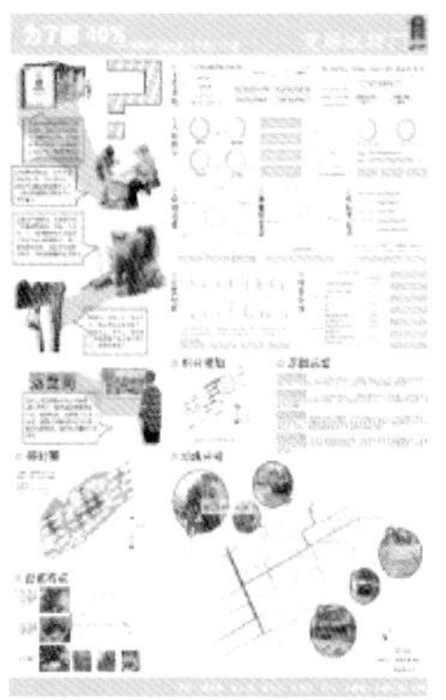

图3-0-32 成果 / Results | 图片来源：编写组自摄

图3-0-33 开营 / Camp opening | 图片来源：编写组自摄

图3-0-34 讨论 / Discussion | 图片来源：编写组自摄

图3-0-35 汇报 / Reporting | 图片来源：编写组自摄

图3-0-36 居民、专家听汇报 / Residents and experts hearing reports | 图片来源：编写组自摄

4 共同缔造：社造联盟创新服务
Co-creation: innovation service by community alliance

社造联盟是什么？
What is community alliance?

- 社区营造联盟，是社区治理的第三方机构，是新生的市场化力量。
- Community alliance, as a third party institution of community governance, is a new marketizing force.

社造联盟有什么优势？
What are the advantages of community alliance?

- 社造机构是社区营造的专业力量，拥有一批专业人才和专业营造手段。
- 作为第三方，可以充分利用市场资源，盘活既有资源，链接政府、居民、市场、社会等多元主体，提供多种专业社区服务。
- Social creation organization is the professional force of community creation, with a group of professional talents and professional creation means.
- As a third party, it can make full use of market resources, revitalize existing resources, link multiple subjects such as government, residents, market and society, and provide a variety of professional community services.

社造联盟能提供什么服务？
What services can the community alliance provide?

- 策划社区活动
- 公益孵化器
- 培育公益人才
- 盘活现有资源
- 引入市场、社会资金
- 提供社会工作
- 提供养老服务
- 教育
- 环保
- 青少年发展
- 扶贫
- 助残
- Planning community activities
- Public welfare incubator
- Fostering public welfare talents
- Revitalizing existing resources
- Introducing market and society funds
- Providing jobs
- Providing old-age care service
- Education
- Environmental protection
- Adolescent development
- Poverty alleviation
- Helping the disabled

上海爱创益公益发展中心：

· 通过组织多种多样的活动，引导居民参与到社区营造当中。

· 针对老人、小孩、青年等不同人群特征开展各种类型活动，广泛吸引居民参与。

Shanghai Aichuangyi Welfare Development Center:

· Guide the residents to participate in the community creation by organizing various activities.

· Carry out various types of activities according to the characteristics of different groups of people, such as the elderly, children and youth, to attract extensive participation by residents.

图3-0-37 社区创意坊·中秋盘绘 / Community creation workshop · Plate drawing in Mid-Autumn Festival | 图片来源：何京洋《创益邻里 营造社区——“爱创益”上海社区更新与社区营造创新实践分享》

图3-0-38 社区创意坊·制作水晶饺 / Community creation workshop · Make dumplings | 图片来源：何京洋《创益邻里 营造社区——“爱创益”上海社区更新与社区营造创新实践分享》

图3-0-39 社区创意坊·制作环保布袋 / Community creation workshop · Make EP cloth bags | 图片来源：何京洋《创益邻里 营造社区——“爱创益”上海社区更新与社区营造创新实践分享》

图3-0-40 社区创意坊·培育植物 / Community creation workshop · Foster plants | 图片来源：何京洋《创益邻里 营造社区——“爱创益”上海社区更新与社区营造创新实践分享》

图3-0-41 社区创意坊·分离种子 / Community creation workshop · Seed separation | 图片来源：何京洋《创益邻里 营造社区——“爱创益”上海社区更新与社区营造创新实践分享》

图3-0-42 社区创意坊·手工制作 / Community creation workshop · Hand work | 图片来源：何京洋《创益邻里 营造社区——“爱创益”上海社区更新与社区营造创新实践分享》

5 保障机制
Guarantee mechanism

建立多元力量共建机制
Establish a multi-force co-construction mechanism

◎ 建立政府部门统筹协调机制

- 建立政府统筹、条块协作的专门工作机制，明确老旧小区改造的相关部门、单位和街道（镇）、社区职责分工。
- 制定工作规则、责任清单和议事规程，形成工作合力，加强项目集成，统筹推进城镇老旧小区改造工作。

◎ Establish an overall coordination mechanism for government departments

- Establish a special working mechanism for government coordination and block collaboration, and clarify the division of responsibilities of relevant departments, units, streets (towns), and communities in the renovation of old communities.
- Formulate work rules, responsibilities lists and procedures to form a joint force of work, strengthen project integration, and coordinate the promotion of the renovation of old communities in cities and towns.

◎ 完善居民参与机制

- 搭建沟通议事平台，畅通居民与社区、街道的沟通渠道。
- 建立线上线下结合的公共信息平台，提供项目进度信息公开、咨询和意见反馈的多种渠道。
- 制定全周期、长效性的参与流程，保障居民的知情权、参与权和监督权。
- 发挥设计师的专业力量，辅导居民有效参与改造，为老旧小区改造提供全过程服务。

◎ Improve residents' participation mechanism

- Build a communication platform to clear the communication channels between residents, communities and streets.
- Establish an online and offline integrated public information platform, and provide multiple channels for project progress information disclosure, consultation and feedback.
- Develop a full-cycle and long-term participation process to protect residents' right to know, participate and supervise.
- Exert the professional strength of designers, help residents effectively participate in the renovation, and provide full-process services for the renovation of old communities.

◎ 健全党建引领机制

- 发挥社区党组织的领导作用，统筹协调社区居民委员会、业主委员会、产权单位、物业服务企业等共同推进改造。组织引导社区内机关、企事业单位积极参与改造。
- 开展小区党组织引领的多种形式基层协商，主动了解居民诉求，促进居民达成共识，发动居民积极参与改造方案制定、配合施工、参与监督和后续管理、评价和反馈小区改造效果等。

◎ Improve the leading mechanism of Party building

- Give full play to the leadership role of the Party organization, and coordinate the community residents' committees, owners' committees, property rights owners, and property service companies to jointly promote the renovation. Organize and guide community agencies, enterprises and institutions to actively participate in the renovation.
- Carry out various forms of primary level consultations led by the Party organization, actively understand residents' demands, promote residents to form a consensus, and mobilize residents to actively participate in the formulation of renovation plans, cooperate with construction, participate in supervision and follow-up management, evaluate and feedback the effects of community renovation.

探索改造资金多方共担机制
Exploring the mechanism of multi-party sharing of funds for renovation

◎ 完善资金分摊规则

· 按照谁受益、谁出资原则，积极推动居民出资参与改造，研究住宅专项维修资金、住房公积金用于城镇老旧小区改造、加装电梯的办法。

· 研究小区范围内公共部分、住宅本体的改造费用出资机制以及财政分类以奖代补机制。

◎ Improve the rules for the allocation of funds

· In accordance with the principle of those who get benefits shall contribute funds, actively promote the residents to participate in and contribute to the renovation, and study the methods of special residential maintenance funds and housing provident funds for the renovation of old urban communities and adding elevators.

· Study the funding mechanism of the renovation cost of the public part and the housing body within the community, as well as the mechanism of replacing the block grant with awards.

◎ 落实居民出资责任

· 动员居民为直接受益或紧密相关的内容出资。根据改造内容产权和使用功能的专属程度制定居民出资标准。

· 居民可通过直接出资、使用住宅专项维修资金、让渡小区公共收益等方式落实出资责任。

◎ Implement the resident's investment responsibilities

· Mobilize residents to contribute funds for directly benefiting or closely related content. Formulate the residents' fund contribution standard according to the degree of exclusiveness of the property rights of the renovation content and the usage functions.

· Residents can fulfill their funding responsibilities by directly contributing funds, applying special residential maintenance funds, and transferring community public benefits.

◎ 加大政府支持力度

· 各级政府分别安排资金支持城镇老旧小区改造。通过一般公共预算、政府型资金、政府债券等渠道落实改造资金。

· 多渠道筹措资金，如住宅用地、商服用地的土地出让收入，提取一定比例作为老旧小区改造专项资金；提取国有住房出售收入存量资金用于城镇老旧小区改造；公共服务设施建设专项资金，优先用于城镇老旧小区改造建设。

◎ Increase government support

· Governments at all levels shall arrange funds to support the renovation of old communities. Funds for renovation are implemented through channels such as general public budgets, government-type funds, and government bonds.

· Raise fund through multiple channels, for example, a certain percentage of the income from the land transfer of residential land and commercial land will be used as special funds for the renovation of old communities; the stock funds of income from the sale of state-owned housing will be used for the renovation of old communities. The special funds for the construction of public service facilities shall be preferentially applied to the renovation of old communities in cities and towns.

5 保障机制
Guarantee mechanism

◎ **吸引市场力量参与**

- 推广政府和社会资本合作（PPP）模式，制定收益约定规则，引导社会资本参与改造。
- 创新老旧小区及小区外相关区域联动改造方式，包括大片区统筹平衡模式、跨片区组合平衡模式、小区内自求平衡模式、政府引导的多元化投入改造模式等。
- 推动专业经营单位参与，明确电力、通信、供水、排水、供气等专业经营单位出资责任。

◎ Attract market forces

- Promote the public-private partnership (PPP) model, formulate the profit treaty rules, and guide social capital to participate in the renovation.
- Innovate the linkage renovation methods of old communities and related areas outside the communities, including large-area overall planning and balance mode, cross-area combination balance mode, self balance mode within the community, and government-guided diversified investment renovation mode, etc.
- Promote the participation of professional business units. Clarify the investment liabilities of professional business units engaged in electricity, communications, water supply, drainage, and gas supply.

◎ **加大金融支持**

- 扶持有条件的国有企业、鼓励引入市场力量作为规模化实施运营主体参与改造，政府注入优质资产，探索项目融资模式。
- 加强政银合作，创新金融服务模式，金融机构为改造项目量身制定融资方案，明确可以未来运营收益作为还款来源，给予开发性金融支持。

◎ Increase financial support

- Support qualified state-owned enterprises, encourage the market forces to participate in the renovation as the main body of large-scale implementation and operations, and the government shall inject high-quality assets and explore the project financing model.
- Strengthen the cooperation between government and banks, innovate financial service models, and the financial institutions formulate tailor-made financing plans for renovation projects, clarify that future operating income can be used as a source of repayment, and grant the development financing support.

◎ **落实税费减免政策**

- 对老旧小区改造免收城市基础设施配套费等各种行政事业性收费和政府性基金。

◎ Implement tax reduction and exemption policies

- For the renovation of old communities, all administrative fees and government funds, such as urban infrastructure supporting fees, will be exempted.

完善配套政策
Improve supporting policies

◎ **建立绿色审查通道**

- 联合审查改造方案。由政府确定的牵头部门，组织相关部门对改造方案进行联合审查，对项目可行性、市政设施和各类技术指标一次性提出审查意见。明确改造项目的优化审批程序，作为改造项目审批及事中事后监管依据。
- 简化立项用地规划许可审批。将项目建议书、可研报告、初步设计及概算进行合并审批。不涉及土地权属变化或规划条件调整的项目无需办理用地许可证。
- 精简工程建设许可和施工许可。不增加建筑面积、不改变建筑结构的改造项目，及符合条件的新建项目，可不办理施工许可证；新建、改扩建项目，合并办理建设许可与施工许可；施工许可和工程质量安全监督手续合并办理；无需办理环境影响评价手续。
- 实行联合竣工验收。由改造项目实施主体组织开展联合竣工验收，简化竣工验收备案材料。

◎ Establish a green review channel

- Joint review of the renovation plan. The lead department determined by the government organizes relevant departments to conduct a joint review of the renovation plan, submit a one-time review opinion on the project feasibility, municipal facilities and various technical indicators, and clarify the optimized approval procedures for renovation projects as the basis for the approval of renovation projects and the operational and post-operational supervision.
- Simplify the approval of land use planning. The project proposal, feasibility study report, preliminary design and budgetary estimate are combined for approval. Projects that do not involve changes in land ownership or adjustments to planning conditions do not need to apply for the land use permit.
- Streamline project construction permits and construction permits. Construction permits are not required for renovation projects that do not increase the building area or change the building structure, and new projects that meet the requirements; for new, renovation and expansion projects, the development permit and the construction permit are integrated; the construction permit and the project quality and safety supervision procedures are integrated, and the environmental impact assessment procedures are not needed.
- Implement joint completion acceptance. The executor of the renovation project organizes the joint completion acceptance and simplifies the completion acceptance filing materials.

5 保障机制
Guarantee mechanism

◎ **制定支持存量资源整合利用政策**

- 整合利用各类公共用房存量资源，用于改建公共服务设施和便民商业服务设施。鼓励机关事业单位、国有企业将老旧小区内或附近的闲置房屋，通过置换、划转、移交使用权等方式交由街道社区统筹。
- 整合利用小区内及小区周边闲置及存量土地，用于建设各类配套和公共服务设施，增加公共活动空间。简化控规调整程序，在符合公共利益与安全的前体下，适度放松各类技术规范要求与技术指标管控。
- 允许将老旧小区存量资产依法授权给项目实施主体开展经营性活动，提供社区便民服务，引导扶持项目实施主体发展成为项目的运营、管理主体。
- 鼓励实施集中连片改造，合理划定改造片区单元，科学编制片区更新规划。对涉及控规调整的，按程序审批后纳入规划成果更新。

◎ Formulate policies to support the integrated utilization of stock resources

- Integrate and utilize various types of public housing stock resources for renovation of public service facilities and convenient commercial service facilities. Encourage government agencies, institutions, and state-owned enterprises to hand over idle houses in or near old communities to the neighborhood communities for overall planning by means of replacement, transfer, or transfer of use rights.
- Integrate the utilization of idle and stock land in and around the community for the construction of various supporting and public service facilities and increase the space for public activities. Simplify procedures for adjustment of regulatory detailed plan, and moderately loosen various technical specifications and technical indicators control under the premise of complying with public interest and safety requirements.
- Allow the stock assets of the old communities to be authorized to the project executor to carry out operational activities, provide community convenience services, and guide and support the project executor to develop into the project operator and manager.
- Encourage the implementation of centralized contiguous renovation, rationally delimit the renovation area units, and scientifically formulate area renewal plans. Those involving the adjustment of regulatory plan shall be included in the planning result update after approval according to procedures.

Construction Guidance for
Renovation of Old Communities in
Jiangsu

江苏老旧小区改造建设导引